THE NEW HEAT THEOREM

THE NEW HEAT THEOREM

ITS FOUNDATIONS IN THEORY AND EXPERIMENT

BY

W. NERNST

TRANSLATED FROM THE SECOND GERMAN EDITION BY
GUY BARR, B.A., D.Sc.

DOVER PUBLICATIONS, INC.
NEW YORK

Published in Canada by General Publishing Company, Ltd., 30 Lesmill Road, Don Mills, Toronto, Ontario.

Published in the United Kingdom by Constable and Company, Ltd., 10 Orange Street, London WC 2.

This Dover edition, first published in 1969, is an unabridged and unaltered republication of the work originally published in 1926. This book is reprinted by special arrangement with Methuen & Co. Ltd., publisher of the first edition.

Standard Book Number: 486-62252-5
Library of Congress Catalog Card Number: 69-15365

Manufactured in the United States of America
Dover Publications, Inc.
180 Varick Street
New York, N. Y. 10014

PREFACE

IN times of trouble and distress, many of the old Greeks and Romans sought consolation in philosophy, and found it. To-day we may well say that there is hardly any science so well adapted as is theoretical physics to divert the mind from the mournful present and to lead it into other spheres; it can, however, offer this attraction only to the few who have studied it for years.

The present booklet deals with an attempt to develop one of the most important chapters of theoretical physics: I refer to thermodynamics, within the scope of which fall not only changes which are physical in the narrow sense, but also, and in particular, chemical processes. Although theory is naturally the origin and aim of our considerations, great weight is laid at the same time on the experimental groundwork.

The new Heat Theorem, which is to be treated here as fully as possible, was first developed by me, and its most important applications indicated in a paper published in January, 1906, in the "Nachrichte der Gesellschaft der Wissenschaften zu Göttingen." The succeeding years brought forth a wealth of further experimental material, but the theoretical nucleus of my theorem underwent, nevertheless, no essential alteration, amplification, or modification, as the reader of this book may readily convince himself. Pollitzer has already given in his excellent book, " Berechnung chemischer Affinitäten nach dem Nernstchen Wärmetheorem " (Enke, Stuttgart, 1912), an almost exhaustive presentation, particularly as regards the practical applications. A new collation appeared desirable, however, not only in view of the more recent theoretical acquisitions which give the new Heat Theorem a widened and surer field of application—I may mention here, in the first place, Debye's T^3-law of specific heats; the theoretical calculation of chemical constants by Sackur, Tetrode, and Stern; and also the later concepts of the so-called "degeneration" of ideal gases at very low temperatures—but also on account of the very large amount of observational material which has since been accumulated.

The collection of the latter has made extensive demands on our laboratories in the last ten years; I hope that the perusal of this book will evoke in my fellow-workers the remembrance of an unclouded, and we hope not unfruitful co-operation. The number of these experimental researches has considerably exceeded a hundred.

It is true that Planck—the first theorist to occupy himself seriously with my Heat Theorem—has given an excellent description of it in the last edition of his " Text-book of Thermodynamics " : the present compilation goes somewhat further in that the theorem is assumed to be valid even for gases of finite density, and, *a fortiori* therefore, also for solutions.

The citations, " Theor. Chem.," which have frequently appeared desirable to make the subject more readily understood, are to the seventh edition of my book (Enke, Stuttgart, 1913).* Both the references to the list, at the end of the book, of

* See Translator's note.

thermodynamical papers from our laboratory and those to the consecutively numbered equations are given by numbers in brackets : I trust this may not lead to any confusion.*

W. NERNST

December, 1917

* In this translation the equations are distinguished by Clarendon numbers.—Tr.

PREFACE TO THE SECOND EDITION

AS the present volume has now been out of print for rather a long time without circumstances rendering it possible for me to revise it completely, I have been obliged to content myself with leaving it practically unaltered and taking account of the recent advances in a Supplement. This has been the more readily possible, as there was no question of any essential changes. Naturally, a number of printer's errors and small mistakes in the old text have been corrected, so that the present reprint should be an improvement on the first edition.

It may be remembered that, in the first edition, the researches adduced to test the new Heat Theorem were principally by my fellow-workers and myself; this has also been the case in the new edition, and it is not entirely without justification. It renders it, however, the more necessary for me to point out that a series of very valuable papers on the subjects which are here in question have appeared from the Physico-chemical Institute of Berlin since my resignation from it; reference is made to these in the Supplement.

W. NERNST

July, 1924

TRANSLATOR'S NOTE

THE second German edition has been closely followed in this translation. Since the Supplement modifies or extends certain matter very appreciably, cross references have been inserted in order to direct attention to the later additions at the appropriate points. One or two later papers have also been cited in footnotes.

The references to Professor Nernst's text-book, " Theoretical Chemistry," follow the pagination of the last English edition (from the eighth-tenth German edition), published by Messrs. Macmillan & Co., Ltd., 1923.

GUY BARR

June, 1926

CONTENTS

	PAGE
PREFACE	v

CHAP.
		PAGE
I.	HISTORICAL INTRODUCTION	1
II.	INVESTIGATION OF GASEOUS EQUILIBRIA AT HIGH TEMPERATURES	15
	1. Transpiration method	15
	2. Method of the heated catalyst	16
	3. Method of the semi-permeable wall	16
	4. Vapour-density determination at high temperatures	16
	5. Explosion method	17
	6. Electro-motive force	18
	7. Calculation of a new equilibrium from other equilibrium measurements	18
	8. Thermal conductivity	19
	9. An example from the dissociation of water vapour	20
III.	SPECIFIC HEATS OF SOLIDS AT VERY LOW TEMPERATURES	24
	1. General	24
	2. The copper calorimeter	26
	3. The vacuum calorimeter: first form	28
	4. Measurement of temperature	34
	5. The vacuum calorimeter: second form	37
	6. Critical	49
IV.	THE LAW OF DULONG AND PETIT	54
	1. General results from our measurements of specific heat	54
	2. Theoretical	57
	3. Debye's T^3-law	61
	4. General formula for the representation of specific heats	61
	5. Calculation of C_v from C_p	65
	6. Determination of atomic weights from specific heats	66
	7. Estimation of the values of ν	67
V.	SPECIFIC HEATS OF GASES	69
	1. General	69
	2. Reduction to the ideal gaseous state	70

CHAP.		PAGE
	3. Experimental methods	71
	4. Application of the quantum theory	72
	5. Monatomic gases	74
	6. Summary	76
VI.	FORMULATION OF THE NEW HEAT THEOREM	78
VII.	PRINCIPLE OF THE UNATTAINABILITY OF THE ABSOLUTE ZERO	87
	1. General	87
	2. Expansion of a solid	90
	3. Chemical change	91
VIII.	SOME IMPORTANT MATHEMATICAL FORMULÆ	94
IX.	APPLICATION OF THE HEAT THEOREM TO CONDENSED SYSTEMS	98
	1. The possibility of an experimental proof of the Heat Theorem	98
	2. Expansion of a solid	100
	3. Melting-point	101
	4. Transition-point	106
	5. Binding of water of crystallization or of hydration	110
	6. Affinity between silver and iodine	113
	7. The Clark cell	116
	8. Investigations of Braune and Koref	118
	9. Summary	119
X.	APPLICATION OF THE HEAT THEOREM TO SYSTEMS CONTAINING A GASEOUS PHASE	121
	1. Statement of the problem	121
	2. Vapour-pressure curves	122
	3. Chemical equilibrium in homogeneous gaseous systems	125
	4. Heterogeneous equilibria	126
XI.	A THERMODYNAMICAL APPROXIMATION FORMULA	128
	1. General	128
	2. Derivation of a vapour-pressure formula	129
	3. Approximation formula for chemical equilibria	133
	4. " Conventional chemical constants "	134
	5. Applications of the simplified approximation formula	139
	6. Cederberg's approximation formula	152
XII.	SOME SPECIAL APPLICATIONS OF THE HEAT THEOREM AND OF THE APPROXIMATION FORMULA DERIVED FROM IT	155
	1. Determination of thermo-chemical data by application of the Heat Theorem to condensed systems	155
	2. Use of the Heat Theorem to control experimental work	157

CONTENTS

CHAP.		PAGE
	3. Electro-chemical applications	158
	4. Application to photo-chemical side-reactions	161
	5. Other applications	165
XIII.	THEORETICAL CALCULATION OF CHEMICAL CONSTANTS	166
	1. Statement of the problem	166
	2. Theoretical calculation of the integration constant of the vapour-pressure formula . . .	169
	3. Experimental test	175
	4. Further applications	187
XIV.	DIRECT APPLICATION OF THE HEAT THEOREM TO GASES	190
	1. Statement of the problem	190
	2. General theory of the degeneration of gases: fundamental assumptions . . .	193
	3. Application of classical thermodynamics .	194
	4. Application of the new Heat Theorem .	195
	5. The physical behaviour of gases at low temperatures	199
	6. A special theory of the degeneration of gases	200
	7. Equation of state for ideal gases at very low temperatures	205
XV.	GENERALIZED TREATMENT OF THE THERMODYNAMICS OF CONDENSED SYSTEMS	211
	1. Formulation of the Second Law for condensed systems	211
	2. Formulation of the new Heat Theorem .	215
	3. Thermodynamic potential	216
	4. Surface tension	217
	5. Coefficient of magnetization . . .	217
	6. Thermo-electricity	217
	7. Introduction of the specific heats . .	218
	8. Use of the T^3-law for specific heats . .	219
	9. Influence of temperature on gravitation .	222
	10. Classification of natural processes . .	225
	11. Recapitulation	226
XVI.	SOME HISTORICAL AND MATERIAL ADDENDA .	227
	1. Early history of the Heat Theorem . .	227
	2. Some further applications of the Heat Theorem to condensed systems . . .	231
	3. Measurement of specific heats . . .	234
	4. Practical applications of the Heat Theorem	234
	5. The Heat Theorem and the Quantum Theory	235
	6. Some questions of principle . . .	237
APPENDIX	. . . .	243
	1. Definitions of symbols and numerical values	243
	2. Tables	245
	I. C_v after Einstein	246
	II. C_v after Debye	247

		PAGE
III. $\frac{E}{T}$ after Einstein	. . . .	248
IV. $\frac{F}{T}$ after Einstein	. . . .	250
V. $\frac{E}{T}$ after Debye .	. . . .	252
VI. $\frac{F}{T}$ after Debye .	. . . .	254
3. List of literature .	. . . . .	256
4. Supplementary notes	. . . .	264
5. Index of authors .	. . . .	279

THE NEW HEAT THEOREM

THE NEW HEAT THEOREM

CHAPTER I

HISTORICAL INTRODUCTION

THE principles of thermodynamics occupy a special place among the laws of Nature. For this there are two reasons: in the first place, their validity is subject only to limitations which, though not, perhaps, of themselves negligibly small, are at any rate minimal in comparison with many other laws of Nature; and in the second place, there is no natural process to which they cannot be applied.

The phenomena of gravitation have hitherto formed an exception to the applicability of the Second Law; but even in this case there are, as we shall see in Chapter XV, possibilities which may be clearly formulated with the aid of the new Heat Theorem.

Until a short while ago there were, as is well known, two thermodynamical principles: (1) The Law of the Conservation of Energy; (2) the Law of the Transmutability of Energy.

According to the first principle there is, for any system, a function U, the negative value of which denotes the content of energy, and which is a simple function of the variables (e.g. volume and temperature) which characterize the system. If we cause the system to undergo any alteration, $U_2 - U_1$ is independent of the route by which we pass from the initial condition having an energy content U_1 to the final condition having an energy content U_2.

According to the second principle there is, for isothermal variations of a system, a function A which has, for such variations, the same properties as U ; $A_2 - A_1$ expresses the maximum external work which can be obtained in the change considered, and this quantity is likewise independent of the nature of the method by which the maximum work considered is obtained.

If then, in the examination of any system, we determine $U_2 - U_1$, or $A_2 - A_1$, by different methods, the invariability of $U_2 - U_1$, as well as that of $A_2 - A_1$, provides us with quantitative relationships between the magnitudes measured.

Let me illustrate this by two examples which appear to be particularly characteristic and are of considerable historical importance :—

(1) Suppose we first mix a definite quantity of water and sulphuric acid at a certain temperature, measure the heat developed and determine the specific heat of the resultant mixture ; again, let us first warm the water and the sulphuric acid by 1°, measure the quantity of heat required, and then observe the heat of mixture at this higher temperature. Then, since the alterations of energy must be equal in the two cases, a relation is supplied between the temperature-coefficient of the heat of mixture and the specific heats of the reacting substances. (Kirchhoff, 1856.)

(2) A concentration cell may do electrical work while at the same time the differences of concentration are equalized : alternatively, differences of concentration may be utilized by means of isothermal distillation to provide external work. If we equate the quantities of work which may be obtained by the two methods, there follows a relation between electro-motive force and vapour pressure. (Helmholtz, 1877.)

From the above-stated laws a further fundamental

relationship may be derived. It has long been recognized that temperature differences may serve to provide external work. More accurately, if a quantity of heat Q flow from a reservoir of heat of absolute temperature $T + dT$ to a second reservoir of temperature T, an amount of work dA may be obtained, by a perfect utilization of the process, which is given by calculation for any suitable method of performing it as

$$dA = Q\frac{dT}{T}.$$

(Carnot, Clausius.) It follows readily from the application of the First Law, if we write U for $U_2 - U_1$ and A for $A_2 - A_1$,

$$dA = (A - U)\frac{dT}{T}$$

or

$$A - U = T\frac{dA}{dT} \qquad . \qquad . \qquad . \qquad \mathbf{(1)}$$

This equation, to which Helmholtz in particular has referred, may be regarded as a summary of the older thermodynamics; but in certain cases the differential

$$S = \frac{dA}{dT} \qquad . \qquad . \qquad . \qquad \mathbf{(2)}$$

(S = Clausius' entropy) requires further explanation, into which we shall enter in Chapter VI: in general, the application of equation **(1)** is quite simple and straightforward, particularly in physical chemistry, and it is done so frequently in recent literature that it may here be taken for granted.

It is well known that Carnot and, especially, Clausius attained their pioneer results by the consideration and calculation of cyclic processes. A cursory examination of the literature of thermodynamics will afford evidence that in every case where a definite solution has resulted, recourse has been had to the discussion of a suitable cyclic process. In considering the Principle of the Unattainability of the

Absolute Zero we, too, shall make use of this fundamental method.

Some questions of an experimental nature may be interpolated here, for it is essential to an understanding of the subsequent historical development to know to what extent one is normally in a position to measure the two quantities A and U which occur in equation (1).

The measurement of U, i.e. of the total change of energy associated with any particular process may, theoretically, always be performed in the following manner. The process in question is allowed to take place inside a calorimeter, and the heat developed or absorbed is measured; any external work given up is calculated to thermal units, so that it may be added to the former amount. In general, the problem may be solved without any particular difficulty and, as far as chemical processes are concerned, there are thermochemical methods available which have been elaborated to a high degree of refinement.

No method of similar universality has hitherto been available for the determination of A, but a suitable experimental arrangement has had to be devised for each case. We shall illustrate this by a few examples, but in many cases, particularly for most organic compounds, there is no possible means of determining the free energy of formation by direct experiment.

If, however, the isothermal change of state considered is purely a function of the volume, such that increase or decrease of the volume causes the change to proceed reversibly in one direction or the other, the problem is very simple: for the work done against the external pressure is the value of A required.

Care must be taken, however, not to regard this method as more general than it really is. If it were desired, for example, to make a theoretical study of the equilibrium between the two crystalline modifications of sulphur according to this method, one would be limited to a very restricted

HISTORICAL INTRODUCTION

temperature interval, if actual facts are not to be ignored. It is true that by means of very high pressures the temperature of equilibrium (transition point) may be noticeably shifted, though the range of temperature is only moderate; but then we are no longer dealing with the modifications of sulphur which are familiar to us, i.e. those occurring under low pressures, such as that of the atmosphere, but with bodies having their properties profoundly influenced by enormous pressure.

In these cases we may frequently make use of the above-mentioned process of isothermal distillation. If we write π_1 and π_2 for the sublimation pressures of the two modifications, we have, neglecting the vanishingly small work given by change of volume multiplied by sublimation pressure,

$$A = RT \log_e \frac{\pi_1}{\pi_2} \qquad . \qquad . \qquad . \quad (3)$$

where A refers to one mol of gaseous sulphur. We may also operate with solubilities, instead of sublimation pressures, by making use of the laws of osmotic pressure.

For gaseous mixtures and solutions, excellent service is frequently rendered to the theoretical calculation by the concept of semi-permeable walls (Gibbs, van't Hoff); the result is thus obtained that the maximum work may be derived from the equilibrium constants of the law of chemical mass action.*

In galvanic cells the progress of the reaction producing the current is determined, according to Faraday's law, purely by the quantity of electricity produced; thus we have here simply

$$A = EF$$

(E = electro-motive force, F = 96540 coulombs). This is one of the most convenient and accurate methods for the determination of A, but it is limited, unfortunately, to special cases.

* Compare in this respect " Theor. Chem.," p. 744.

For the conversion of thermal into radiant energy, the pressure of light plays a rôle analogous to the vapour pressure of solid bodies.

The most general method for the determination of A is the particular theme of this monograph : the Heat Theorem which is here to be treated allows us to calculate A from U, if we know U at any one temperature and also the heat capacity of the system down to the absolute zero before and after the change considered. Obviously we may, *vice versa*, also find U if A is measured at any temperature and the specific heats down to zero are known.

From the standpoint of classical thermodynamics, it may be asked—Is equation (1) the last word in thermodynamics ? If not, it can only be that there is a closer relationship between the two thermal functions A and U than is already given by equation (1).

Speculations on this subject have in fact been several times put forward in the consideration of chemical and electro-chemical processes, where people did not quite realize that any extension of equation (1) must be of a general nature, i.e. could not be restricted to chemical processes.

The first conscious attempt in this direction is due to Julius Thomsen, who repeatedly stated as early as 1852, in his " Contributions to a System of Thermo-chemistry," that vigorous manifestations of chemical affinity were accompanied by vigorous development of heat, and that chemical processes associated with an absorption of heat were of rare occurrence. Hence he arrived at the following conclusion : " When a body falls it develops a certain mechanical effect which is related to its weight and to the distance traversed. In chemical reactions which take place in their normal direction, a certain ' effect ' is again produced, but in this case it appears as heat. In the development of heat we have a measure of the chemical force developed in the reaction."

We shall see later that chemical processes cannot, at

HISTORICAL INTRODUCTION

any rate above the absolute zero, be regarded as phenomena of attraction comparable with the falling of a stone. But small blame attaches to Thomsen in this respect, seeing that the same method of treatment continues to be frequently attempted even to-day, in spite of the kinetic theory of heat. As a matter of fact, Thomsen himself recognized the untenability of his conceptions as early as 1870, impressed though he was by the results of his extremely ingenious method of determining the affinity between acids and bases.

The same law was propounded in 1869 by Berthelot, the other father of thermo-chemistry, and was vigorously supported by him for many years. Berthelot's formulation runs thus :—

" Every chemical transformation which takes place without the intervention of external energy tends towards the production of that substance, or system of substances, which will give the greatest development of heat."

Both formulations, the earlier one of Thomsen and the later one of Berthelot, are equivalent to writing equation **(1)** for all temperatures as

$$A = U.$$

It is unnecessary to give a detailed proof here of the inadmissibility of this rule, but reference to an observation of Horstmann's will be of use to illustrate it. He remarks that the recognition of a chemical equilibrium, or, which comes to the same thing, of a reversible reaction, suffices to refute Berthelot's principle. For in the neighbourhood of the chemical equilibrium the reaction takes place in one or the other sense, depending on the proportions of the reacting components, according as we are on one or the other side of the equilibrium : hence the reaction must proceed in the one case, in agreement with Berthelot's principle, with evolution of heat, but in the other, in contradiction to the principle, with absorption of heat.

We have already pointed out that the electro-motive force of a galvanic cell is proportional to the affinity of the process which yields the current. The Thomsen-Berthelot principle may therefore be expressed by saying that the electro-motive force of the galvanic cell must be proportional to the heat evolution per electro-chemical gramme-equivalent. It is of historical interest to note that this rule occurs, albeit only *en passant*, in Helmholtz's famous work on the conservation of energy (1847). The method of calculation there indicated was later carried out, for several examples, by William Thomson. More precise test has shown, in agreement with the above considerations, that the electro-motive force of galvanic cells can in fact be frequently calculated with great accuracy from the heat developed, particularly when vigorous affinities are concerned, but that the rule cannot be called by any means rigid.

For a certain class of reactions, namely, for the case in which a gas is produced from one or more solid bodies, le Chatelier (1887), and later Matignon and Forcrand, found the following approximate relationship ; if Q denote the heat developed at constant pressure, T' the absolute temperature at which the dissociation pressure p of the gas produced is equal to atmospheric pressure, then

$$\frac{Q}{T'} = ca.\ 33.$$

The Second Law gives in this case, if we neglect the variability of Q with the temperature,

$$\log_e p = -\frac{Q}{RT} + \text{const.}$$

We see at once that the le Chatelier-Matignon rule gives a value about 33 for R times the undetermined constant of integration. The rule is only approximately true, but it does give us a valuable indication, and it deserves more attention than was formerly given to it. We shall learn later how to formulate the rule more accurately.

In 1904 van't Hoff enunciated a somewhat unsatisfactory

HISTORICAL INTRODUCTION

hypothesis. If the influence of temperature on U be represented by the (apparently) simplest law,

$$U = U_0 + aT, \qquad . \quad . \quad . \quad (4)$$

integration of **(1)** gives

$$A = U_0 + a_0 T - aT \log_e T, \qquad . \quad . \quad (5)$$

where a_0 signifies the constant of integration; since formula **(1)** is a differential equation, any integration of it will of course introduce a constant which, though independent of temperature, varies with the nature of the system considered, and is therefore unknown at first. Now van't Hoff assumed a_0 to be small. This hypothesis is, however, not only arbitrary but apparently inexact in itself. For if we take even the case where a_0 is equal to zero, we should need only to alter the scale of temperature, e.g. to divide the interval between the melting-point and boiling-point of water into a million instead of into a hundred parts, in order to obtain a finite, even a large, value for a_0. It is hardly probable that the laws of nature should be such as to make Celsius divide this temperature interval into a hundred parts and happen to choose water as his normal substance.

The older attempts to extend equation **(1)** miscarried therefore; but the problem was at any rate sharply defined. Next to Helmholtz, le Chatelier gave * the clearest statement of it in 1888. I may reproduce his words:—

"It is very probable that the constant of integration, like the other coefficients of the differential equation, is a definite function of certain physical properties of the reacting substances. Determination of the nature of the function would lead to a full knowledge of the laws of equilibrium. It would determine, *a priori* and independently of new experimental data, the complete conditions of equilibrium corresponding with a given chemical reaction. It has hitherto been impossible to determine the exact nature of this constant."

* "Les équilibres chimiques," Paris, 1888, p. 184.

I myself have also given attention to the problem attacked in the above-mentioned investigations. In the first edition of my text-book, "Theoretical Chemistry" (1893), I observed, in dealing with Berthelot's rule, that although the rule could not be accepted as a law of nature, yet it was surprisingly often useful in practice, and "that it might well be possible that Berthelot's principle in a clearer form should one day again be found to hold." In particular, I directed attention to what had already been noted by Horstmann, that the affinity and the development of heat in a reaction were often practically identical, especially for solids.

I have pointed out earlier * that the formula

$$A - U = T \frac{dA}{dT} \quad . \quad . \quad . \quad . \quad (1)$$

showed two limiting cases as special simplifications, namely,

(1) the special case $U = 0$, $A = T \dfrac{dA}{dT}$,

i.e. A is proportional to the absolute temperature;

(2) the special case $A = U$, i.e. $\dfrac{dA}{dT} = 0$,

or A, and hence also U, is independent of the temperature.

The former case is realized in nature in the expansion of ideal gases and the mixture of dilute solutions; the most important temperature measuring instrument, the air thermometer, depends upon this fact.

The second case is found in all systems in which only gravitational, electrical, or magnetic forces are operative, and then the behaviour can be represented, of course, by a force function independent of temperature, the potential. The statement of the above-mentioned rules of Helmholtz and Berthelot shows that this case is at least partially realized in electro-chemical and purely chemical processes. Finally, I showed in 1894 that, again in certain cases, this

* "Theor. Chem.," 2nd ed., p. 37 (1898).

HISTORICAL INTRODUCTION

behaviour is to be observed on mixing concentrated solutions and that it occasions a marked simplification there.

In the experimental direction it seemed to me very desirable to apply the fundamental equation **(1)** over a range of temperature as large as possible ; this can be most completely effected by the examination of chemical equilibria in gaseous systems, where the field is limited in the upper direction only by the possibility of carrying out measurements at sufficiently high temperatures.

These researches, to which a short reference will be made in the following chapter, did not lead immediately to the solution of the problem, although they afforded me many *points d'appui*. But relying on the available, though only indirectly evidential material and assisted by a lucky chance, I arrived at the following conception.

Although at very high temperatures the ideal gaseous state, i.e. the case $U = 0$ (compare above, p. 10) prevails in gaseous systems, in spite of certain disturbances such as dissociation, every system, on the contrary, given sufficiently strong cooling, i.e. sufficiently near the absolute zero, converges towards the second limiting case $A = U$.

It is well known how great a rôle graphical representation, e.g. of cyclic processes, has played in the development of thermodynamics. This is not accidental, but lies in the essence of the thermodynamic method of treatment. For no other assumption is here made as to the nature of the functions involved, e.g. as to the manner in which vapour pressure varies with temperature, than that the functions employed are continuous and continuously variable. Graphical representation is thus more general than, for example, development according to whole powers of absolute temperature, which is equally available in the thermodynamical treatment of such functions. We shall therefore make use of graphical representation to illustrate the above hypothesis.

First, let us consider the behaviour of the quantity A in

the sense of van't Hoff's rule **(4)** and the resulting integration **(5)**. Since the integration constant a_0 is undetermined, A is not definitely fixed, but we obtain a family of curves in which each of the infinite number of curves with the fundamental equation **(1)** is included (Fig. 1). We see further that, for T = 0, A becomes a tangent to the ordinate, i.e.

$$\lim \frac{dA}{dT} = \infty \text{ (for T = 0)} \quad . \quad . \quad . \quad \textbf{(6)}$$

If, on the other hand, we introduce the new hypothesis,

A = U (for small values of T),

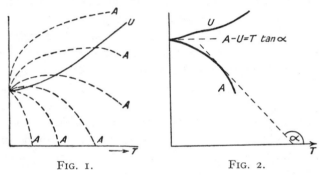

FIG. 1. FIG. 2.

it follows that A and U must coincide at low temperatures, that is, we must have

$$\lim \frac{dA}{dT} = \lim \frac{dU}{dT} = 0 \text{ (for T = 0)} \quad . \quad . \quad \textbf{(7)}$$

The hypothesis selected by van't Hoff as apparently the simplest is hence impossible, for at low temperatures U must become parallel to the abscissa. The strange conclusion **(6)** also disappears, and we obtain the simple result that A is now definitely fixed if U is known as a function of the temperature down to the absolute zero. In fact, if curve U (Fig. 2) is given, we have to identify A at the beginning of its course with U, on the terms of our hypothesis; the fundamental equation **(1)**,

$$\frac{dA}{dT} = \tan \alpha = \frac{A - U}{T},$$

now gives us the subsequent course of the A curve definitely, since at any point we know in what direction we have to extend A.

Thus we have found the general solution of the problem which Thomsen and Berthelot propounded, and this not only for chemical processes, though that part of the question is particularly important, but for every natural occurrence.

Equations (7), which form the essential content of my Heat Theorem, are given in a paper (1) published in full in 1906; I also referred there to a number of possibilities for their application which depended partly, it is true, on account of insufficient experimental support, on formulæ which were approximate only, but were expressly mentioned by me as being so. I noted further that, for the immediate application of equations (7), systems in which gases played a part were to be excluded, since we had no idea what became of a gas if we cooled it continuously, i.e. avoiding condensation, down to the absolute zero. For practical applications it was therefore necessary to deal at very low temperatures with condensed systems, and to make the change to the gaseous state at high temperatures by introducing the vapour pressure. If this calculation is performed in general, one is led to " chemical constants."

Just recently the " degeneration of gases " foreseen by me from the quantum theory has filled this gap and allowed the immediate application of my Theorem even to gases: thus chemical constants have been referred to thermal quantities, namely, specific heats.

In this way the general solution has been obtained to the problem which I originally propounded, namely, the calculation of the maximum work (free energy, chemical affinity, electro-motive force, vapour pressure, etc.) from purely thermal quantities, namely, specific heats, heats of transformation, heats of reaction of all sorts; or, expressed in more general terms, whereas before U could be calculated when A was known for all temperatures but not the converse, the latter is now also possible.

As has already been mentioned, my paper of the year 1906 contained my Heat Theorem in the form which it has retained to the present day; nor, I think, will the future produce any changes, for the above paper gave the essential solution.

In later years I was concerned particularly in broadening the experimental foundations, and it was to be foreseen that such work would lead to many subsidiary results.

In fact, new points of view and new methods had to be discovered for the examination of chemical equilibria. Of late years experimental knowledge has been extended, especially in the field of specific heats, both of solid and gaseous bodies, and as a result of this work theory has made great advances, expressed particularly in Debye's T^3-law, and in the complete elucidation of the time-honoured law of Dulong and Petit.

The following chapters (II to V) are devoted to these researches.

Like the two Laws of classical thermodynamics, my Heat Theorem is the quintessence of a large amount of experimental material, some of a positive and some of a purely negative character. Theoretical arguments which operate with more or less arbitrary hypotheses can only give support of a secondary character, and in particular, though many eminent theorists have overlooked it, considerations of molecular theory are less suited to correct established thermodynamical laws than to be themselves accommodated thereto. It appears useful, however, to put on one side the principle of the impossibility of a *perpetuum mobile* (First Law of Thermodynamics), the principle of the impossibility of a heat engine which continually transforms the heat of its surroundings into external work (Second Law), and the principle of the unattainability of the absolute zero (the new Heat Theorm); we shall go into them more closely in Chapter VII.

CHAPTER II

INVESTIGATION OF GASEOUS EQUILIBRIA AT HIGH TEMPERATURES

THESE researches, which I began in the Göttingen Laboratory, and which afforded a test of the Second Law over a very large interval of temperature, are not so closely related to the special subject of this book as to require more than a concise review. I shall therefore content myself with a short description of the different methods, some new and some improved, which we used in the most important equilibrium investigated by us, namely, the dissociation of water vapour.

1. Transpiration Method.—The gas or mixture of gases under investigation is allowed to flow into an enclosure heated to a uniform temperature, and to escape through a narrow capillary. If the conditions are suitably selected equilibrium is practically completely attained in the heated space, and is not disturbed during the efflux, so that analysis of the cooled gas gives the chemical equilibrium. I have developed the theory of this method as regards its use in determining the equilibrium in the formation of nitric oxide from air.* As was shown by Knietsch, in his work on the formation of sulphur trioxide (1901), the attainment of equilibrium may be accelerated by the introduction of catalysts into the heated space: there is then, however, the danger that particles of the catalyst, albeit only in minimal quantity, may enter the capillary exit and there subse-

* " Zeitschr. anorg. Chem.," **49,** 213 (1906).

quently displace the equilibrium : this is a source of error to which sufficient attention has not, I think, been paid hitherto.

2. Method of the Heated Catalyst.—If a heated catalyst is introduced into a gas, in which the reaction velocity is negligibly small in virtue of the low temperature prevailing, chemical equilibrium is set up in the immediate neighbourhood of the catalyst; after some time the composition corresponding with equilibrium at the temperature of the catalyst will be established, by diffusion and convection, throughout the whole of the gas. In the case of water vapour, one may use as catalyst simply a platinum wire, the temperature of which is determined from its resistance; Langmuir,* on my recommendation, has used this method with good results.

3. Method of the Semi-permeable Wall.—It is well known that platinum is permeable by hydrogen at high temperatures. If, therefore, an evacuated platinum bulb be exposed to strongly heated water vapour, that pressure of hydrogen will forthwith be established in it which corresponds with the concentration of hydrogen set free by dissociation of the water; this was recognized by Löwenstein,† and exactly demonstrated in my laboratory at Göttingen. For very high temperatures iridium may be used instead of platinum, as was found by v. Wartenberg.‡ This extremely accurate method is at present limited to those cases in which hydrogen takes part in the equilibrium; Preuner (9) used the same method to investigate the dissociation of sulphuretted hydrogen.

4. Vapour-density Determination at Very High Temperatures.—Victor Meyer was able to extend up to $1700°$ the method of air displacement discovered by him, and thus to

---

* " Dissertation," Göttingen, 1906; " Am. chem. Soc.," **28**, 1357 (1906).

† " Zeitschr. physik. Chem.," **54**, 707 (1906).

‡ Ibid., **56**, 526 (1906).

determine in many cases whether dissociation occurred to any marked extent at the temperature concerned. The originally somewhat tedious method may be much simplified by the use of electric furnaces, and especially by reduction of the dimensions : I have been able, by the use of small iridium bulbs, to make measurements * of sufficient accuracy up to above 2000°. This method is extremely convenient for deciding whether dissociation occurs, but is not usually applicable for exact quantitative measurements.†

5. Explosion Method.—This method has already permitted us to carry out measurements up to nearly 3000°, and we hope to go considerably higher by further development of it. The required high temperatures are attained by explosion of a gaseous mixture in a closed bomb, and there is then the advantage that no vessels of refractory material are required. Measurements of the maximum pressures of explosion were undertaken by different authors some time ago, but the older determinations were very considerably in error on account of inertia of the manometer employed. It was not until after the protracted endeavours of Dr. Pier (24) and (42), my pupil, and for several years my able assistant, that it was possible to measure the true course of the pressure and, in particular, the maximum pressure, with very great accuracy. He used a membrane of high frequency, the deflections of which resulting from the explosion pressure were photographically recorded on a time basis with the aid of a small mirror. Since the maximum pressure will vary according to the mode of attainment of chemical equilibrium at the maximum temperature of the explosion, the pressure measurement allows conclusions to be drawn concerning the equilibrium ; at the same time, the specific heats of the reacting gases are found by this means with relatively great accuracy, and that up to otherwise inaccessible temperatures.

* " Theor. Chem.," p. 299.
† Cf. ibid., p. 531, and Löwenstein, *loc. cit.*

In certain cases a difficulty occurs in the calculation of the results: this consists in the formation of new and unknown endothermic compounds, an occurrence which, though very interesting, is difficult to take into account. It would therefore constitute a great advance if the maximum temperature of the explosion, in addition to the maximum pressure, could be measured with the necessary accuracy. Experiments with this object have, however, not yet led to any satisfactory results; possibly it will be easier to make use of the temperature of freely burning flames to obtain supplementary assistance.

6. Electro-motive Force.—Helmholtz showed, as early as 1889, that the dissociation of water could be theoretically calculated from the electro-motive force of polarization, but in this particular case the counter E.M.F. of the reversible decomposition of water cannot be directly determined. In other cases, however, the method has proved itself applicable; e.g. from the E.M.F. of the hydrogen-chlorine gas cell and the HCl-vapour tension of the acid used, the dissociation of hydrochloric acid can be determined at ordinary temperatures, and then that at very high temperatures obtained with the aid of the molecular heats of the reacting gases, which have been measured by Pier up to very high temperatures (25).

7. Calculation of a New Equilibrium from Other Equilibrium Measurements.—This method is capable of very numerous applications. My co-workers and I determined the equilibrium of the two reactions—

$$N_2 + O_2 = 2NO,$$
$$2H_2 + O_2 = 2H_2O;$$

from these the action of nitrogen on steam at very high temperatures, in the sense of the equation

$$N_2 + 2H_2O = 2NO + 2H_2$$

may be calculated; Tower,* at my suggestion, checked the

---

* "Ber. d. chem. Ges.," **28,** 2946 (1905).

results of the calculation. At the high temperatures concerned, ammonia is no longer appreciably stable, and is therefore not formed.

In connection with the dissociation of water vapour, reference may be made also to the following application The dissociation of hydrochloric acid,

$$H_2 + Cl_2 = 2HCl,$$

has been established with considerable accuracy according to method 6, and also in other ways. The equilibrium of the reaction

$$2Cl_2 + 2H_2O = 4HCl + O_2$$

(Deacon process) has been accurately determined by my pupil Vogel von Falkenstein, and others. By the combination of these results the dissociation of water vapour is obtained (25).

8. Thermal Conductivity.—In a homogeneous gaseous system the conductivity, as I deduced * from theory, must assume unusually high values if there is an equilibrium being rapidly established in it, the disturbance of which is associated with great development of heat. R. Goldschmidt was able to show, in my laboratory at Göttingen as early as 1901, that dissociation could be recognized qualitatively by this means at very high temperatures inaccessible at that time to accurate measurement. More recently, at v. Wartenberg's suggestion, Stafford [†] attempted in our laboratory to determine the different stages of dissociation of sulphur by this method. The basis for the quantitative calculation of the dissociation by this means is given in my paper above cited, of which Langmuir has also recently made use.

Recently Isnardi (102*a*) has brought the process to a high degree of perfection and applied it, as did Langmuir

---

* " Festschrift Boltzmann," p. 904 (1904).
† " Zeitschr. physik. Chem.," **77**, 66 (1911).

before him, to the dissociation of hydrogen; we shall return to these results again in Chapter XIII.

9. An Example from the Dissociation of Water Vapour.—As an example of the application of the above methods, I may here cite our results on the dissociation of water vapour. The heat of combustion of hydrogen is well known, and the specific heats required have been measured with sufficient accuracy by Pier and his followers by the explosion method. The Second Law may therefore be applied, as in hardly any other case, over a range of temperature in which there are no gaps.

TABLE I

Extent x of Dissociation of Water Vapour

T.	100x obs.	100x calc.	Method.	Observer.
290	$0.46 - 0.48 \times 10^{-25}$	0.466×10^{-25}	6	Lewis, Brönsted.
700	7.6×10^{-9}	5.4×10^{-9}	7	Nernst.
1300	0.0027	0.0029	2	Langmuir.
1397	0.0078	0.0085	1	Nernst and v. Wartenberg.
1480	0.0189	0.0186	1	Nernst and v. Wartenberg.
1500	0.0197	0.0221	2	Langmuir.
1561	0.034	0.0369	1	Nernst and v. Wartenberg.
1705	0.102	0.107	3	Löwenstein.
2155	1.18	1.18	3	v. Wartenberg.
2257	1.77	1.76	3	,,
2337	2.8	2.7	5	Bjerrum, Siegel.
2505	4.5	4.1	5	,, ,,
2648	6.2	6.6	5	,, ,,
2731	8.2	7.4	5	,, ,,
3092	13.0	15.4	5	,, ,,

The numbers given under "calc." are taken from the paper by Siegel (97), who has repeated the calculations made by v. Wartenberg and myself (1906) in the light of the latest measurements of specific heats.

For the mean specific heats between 0° and T° we take—

INVESTIGATION OF GASEOUS EQUILIBRIA

$$C_{vH_2}^{(0,T)} = 4.650 + 3.75 \times 10^{-4}T \Big] *$$
$$C_{vO_2}^{(0,T)} = 4.850 + 3.75 \times 10^{-4}T \Big]$$
$$C_{vH_2O}^{(0,T)} = 5.750 + 0.783 \times 10^{-3}T - 0.626 \times 10^{-6}T^2$$
$$+ 4.56 \times 10^{-10}T^3 - 2.18 \times 10^{-17}T^5.†$$

These formulæ rest on explosion experiments covering an interval of 273° to 2700° for the first two gases, and of 273° to 2900° for the last. They also reproduce excellently all the other available trustworthy measurements for the range of temperature from 273° to 1600°. The extrapolation by some hundreds of degrees upwards may be made without any hesitation, as no sudden changes are to be expected here, according to all other experience.

On the other hand, the above formulæ may on no account be extrapolated downwards *ad libitum* (cf. Chapter IX). It must be emphasized that the region of their validity, and hence also that of the equation given below, ends somewhere about $T = 200°$.

The heat of formation of a molecule of water vapour at 290° abs. at constant volume amounts to 57,290 cals.,‡ or 114,580 cals. for the reaction $2H_2 + O_2 = 2H_2O + U_T$. Then, according to the First Law,

$$U_T = U_0 + T(2C_{vH_2}^{(0,T)} + C_{vO_2}^{(0,T)} - 2C_{vH_2O}^{(0,T)}),$$

and, by making use of the above numerical values,

$$U_T = 113820 + 2.65T - 4.41 \times 10^{-4}T^2 + 1.252 \times 10^{-6}T^3$$
$$- 9.12 \times 10^{-10}T^4 + 4.36 \times 10^{-17}T^6.$$

If this expression is introduced into the equation of the reaction isochore

$$U_T = \frac{d \log_e k}{dT} RT^2,$$

* Calculated to absolute temperatures from the formulæ given by Pier, " Z. f. Elektroch.," **15**, 539 (1909).
† According to Siegel.
‡ See Pier (24).

and we integrate and convert into logarithms to the base 10, we find

$$\log_{10} k = -\frac{24900}{T} + 1{\cdot}335 \log_{10} T - 0{\cdot}965 \times 10^{-4}T + 0{\cdot}137 \times 10^{-6}T^2 - 0{\cdot}665 \times 10^{-10}T^3 + 0{\cdot}1007 \times 10^{-17}T^5 + J.$$

Taking the minimum numbers obtained by v. Wartenberg and me (3) by the transpiration method, which constituted the first accurate determinations of the dissociation of water vapour, and may be still the most reliable, we have

$$J = -1{\cdot}08.$$

This value has been used to calculate the numbers given in the foregoing table : we see that the above thermodynamical equation embraces a large interval of temperature, and reproduces extremely well the observed percentage degrees of dissociation, varying from 10^{-25} to 10^{+1}. This is evidence that the very varied methods which have been used in the case of the reaction

$$2H_2O = 2H_2 + O_2$$

have given reliable results. The same may well be assumed for the gaseous reactions which we have studied in which the same methods have been used, though not, naturally, in the same variety in each case

It may be remarked, incidentally, that the second value in the above table has been calculated in the following manner (25), which is not without interest.

The electro-motive force of the hydrogen-chlorine gas cell is well known ; from this the dissociation of hydrochloric acid gas may be derived, knowing the HCl-vapour pressure of the solutions used. With the further assistance of the equilibrium in the Deacon process,

$$O_2 + 4HCl = 2H_2O + 2Cl,$$

which has been recently examined in our laboratory by Vogel v. Falkenstein, the dissociation of water vapour may

INVESTIGATION OF GASEOUS EQUILIBRIA

now be calculated. The relatively insignificant discrepancy, which is encountered in this case in the above table, indicates that there are still corrections to be applied somewhere in the above-mentioned foundations of the calculation; but there can be only unimportant experimental errors. In any case, the successful thermodynamical correlation of measurements of absolutely different type is remarkable.

Of particular importance is the result of a similar calculation, namely, the derivation of the electro-motive force of the oxy-hydrogen cell from the dissociation of water vapour. Calculation, the principle of which dates back to Helmholtz, led v. Wartenberg and me (1906) to the result

$$E = 1 \cdot 2322 \text{ volts at } T = 290°:$$

the direct measurements of the hydrogen-oxygen cell had given a considerably smaller value, of some 1·18 volts, but we were even then able to show that there was apparently no doubt of the correctness of our figure. Soon afterwards both Lewis and Brönsted confirmed our value by indirect potential measurements * (replacement of the oxygen electrode by Ag_2O or HgO electrodes); the mean of these measurements gives

$$E = 1 \cdot 231 \pm 0 \cdot 003 \text{ volts.}$$

The extent of dissociation given in the first column of the above table as "obs." is derived from this figure; that given under "calc." results from the above thermodynamical formula, and leads to the value

$$E = 1 \cdot 2325 \text{ volts,}$$

which is, of course, practically identical with that just given as calculated earlier by v. Wartenberg and myself.

---

* Cf. "Theor. Chem.," p. 850. For the sake of completeness I have included these measurements in the table on p. 20, although they did not come from my laboratory.

CHAPTER III

SPECIFIC HEATS OF SOLIDS AT VERY LOW TEMPERATURES

1. General.—The equations (7) which constitute the substance of my Heat Theorem, at once suggested a new experimental problem, namely, the determination of specific heats down to temperatures as low as possible. It is only if the specific heats are known that we can quite safely calculate the quantities A and U at low temperatures by means of the two already known laws, and so test the new Heat Theorem; for these quantities are, for many reasons, practically always inaccessible to direct measurement at low temperatures.

At low temperatures, of course, the difference between " solid " and " liquid " becomes obliterated in many respects; solids like quartz glass and ordinary glass are, for instance, to be regarded as supercooled liquids, although, like crystalline solids, they possess elasticity to resist deformation (Tammann). It seems quite possible that a knowledge of the energy content of supercooled liquids in general may be obtained either by elaborating the technique of super-cooling or by some indirect means such as by examining adsorbed liquids. With rarefied gases the difficulties are augmented by the apparently prohibitive fact that extensive super-cooling below the condensation point is never possible; but in this case there are more hopes (cf. Chap. XIV) of filling up the gap by theoretical work.

I therefore set myself to perfect methods for the determination of the specific heats of solids at very low temperatures, and I think I may say that I have solved the problem.

SPECIFIC HEATS OF SOLIDS

We shall see in the next chapter that the results obtained have given a decisive bent to our conceptions of the energy content of solids, and hence also of the theory of the solid state in general.

Thanks to the work of Behn, Tilden, Dewar, and others, something was already known of the behaviour of specific heats at low temperatures; but there was no method which would give, not merely the mean energy content over a considerable temperature interval, but the true specific heat down to the lowest possible temperatures.

From what was known, it was by no means certain even that the problem was soluble. It was conceivable that at low temperatures phenomena of retardation caused, say, by a large decrease in conductivity might have made measurements impossible. Such difficulties did not arise. As the investigations of Eucken (82) in particular show, the conductivity has, on the contrary, a tendency to increase considerably at low temperatures. Since also the thermal capacity falls rapidly at the same time, temperature differences are very quickly equalized at low temperatures, a circumstance which is extremely favourable to accuracy in calorimetry.*

In conjunction with my fellow-workers, I constructed two types of calorimeter, the copper calorimeter and the vacuum calorimeter: since the latter apparatus is of particular importance for our purpose, it will be described in more detail in its different forms.

* It is of interest that at low temperature the flow of heat becomes more and more comparable with the flow of electricity in good conductors; if, owing to vigorous cooling, all gases have disappeared and radiation is extremely small, heat may be conducted almost quantitatively through a copper wire, for example, and the thermal conductivity measured in a Wheatstone bridge. Since, in all probability, heat has inertia, it is possible that at very low temperatures, with the resulting high conductivity, an oscillatory discharge of thermal differences of potential might occur under certain circumstances.

2. The Copper Calorimeter.

The first description of this apparatus was given by W. Nernst, F. Koref, and F. A. Lindemann (36): detailed instructions for its use were given later by Koref (60), who has carried out a large number of valuable measurements with it. The work was then continued by A. S. Russell (79a) and Ewald (99): the latter, in the examination of ammonium compounds, discovered the remarkable fact that there is here a transient decrease of specific heat with rise of temperature over a certain range, which points to internal rearrangements in the ammonium radicle.

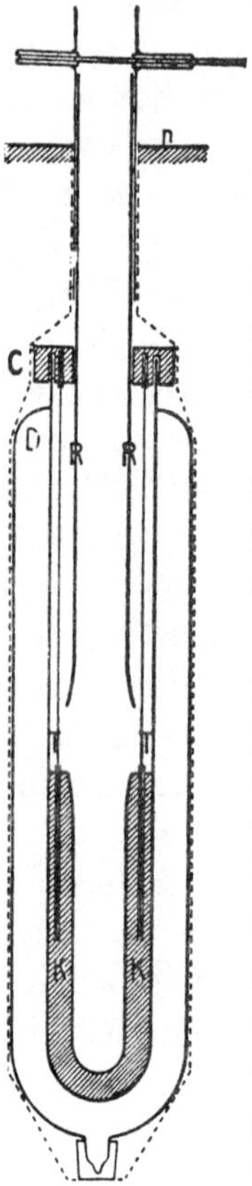

The calorimeters devised by Schottky (22) and Schlesinger (20) may be mentioned as forerunners of the copper calorimeter in certain respects.

The copper calorimeter works on the principle of the "mixture" calorimeter: in place of the calorimetric liquid there is a copper block with good thermal insulation, provided with an axial cavity for receiving the warmer or cooler substance under investigation. The change in the temperature of the block is measured with the aid of thermocouples; the good thermal conductivity of copper ensures that all parts of it are always at practically the same temperature. With a view to the best thermal insulation, it is contained in a double-walled vessel which is evacuated and silvered.

FIG. 3.

The arrangement of the complete apparatus is shown in Fig. 3. K is the copper block weighing some 400 gms., which is set in the vacuum vessel D; TT are the thermocouples. The lower junctions are contained in small thin-walled glass tubes inserted into the copper block; to pro-

SPECIFIC HEATS OF SOLIDS

mote the exchange of heat, the lower halves of the glass tubes are filled with Woods' metal, which also surrounds them on the outside. The other junctions are in a second copper block of annular shape which forms the stopper at the top of the vacuum vessel. Through the opening in this passes a glass tube R, serving for the introduction of the substance under examination : the upper end of it can be closed with a cork.

The whole apparatus is immersed up to n in a bath of constant temperature ; ice or carbon dioxide snow is usually employed for the purpose. The temperature of the upper copper block is thus maintained constant. Since there are difficulties in making the joints tight, the whole of the apparatus is encased watertight by a sheath of thin copper foil suitably soldered ; this is shown dotted in the figure.

The variation of the temperature of the copper block K, relative to that of the block C, caused by the introduction of the substance, is determined by means of ten copper-constantan couples in series, the electro-motive force being measured by means of a Siemens and Halske millivoltmeter. As these instruments allow very accurate readings to be taken, using a lens if necessary, it was possible to employ one of them instead of a mirror galvanometer.

The glass wall which surrounds the calorimeter block is naturally made as thin as possible, and it is desirable that there should be good thermal contact between them. Otherwise irregularities in the rate of change of temperature may easily be observed. The block was therefore cemented with Woods' metal into the vacuum vessel. For this purpose some of the alloy was put into the vessel ; the block, electrically heated from the inside, was then introduced and pushed down a suitable depth into the Woods' metal.

The substances under examination were contained in thin-walled silver vessels, weighing only a few grammes, and consequently possessing only a moderate capacity. The silver vessels were a good fit in the bore of the calorimeter.

To enable the temperature to be accurately determined before they were introduced into the calorimeter, a thermocouple was pushed into a little silver tube soldered into the middle of the vessel. When necessary (for example, in the case of calcium oxide) the vessels were hermetically sealed by soldering the cover on. In some few cases (e.g. mercury and bromine) the substance was first enclosed in a small thin-walled glass tube which was sealed.

Rapid equalization of temperature may be further promoted by keeping the bore of the copper block filled with hydrogen instead of with air.

As regards the arrangements for bringing the substance to be introduced into the copper block to a known higher or lower temperature, the above-cited work of Koref, and the other literature mentioned, may be consulted. Once the apparatus is properly set up, results may be obtained with it rapidly and very accurately. It should be useful, and in most cases preferable to the ice calorimeter, not only for the determination of heat capacities, but whenever the calorimetric determination of small quantities is required.*

On the other hand, it gives the specific heat only as a mean value over a more or less extended interval of temperature. At very low temperatures, e.g. at the temperature of boiling hydrogen, there might be difficulties in its use; this has not been tried. The apparatus next to be described is much to be preferred in these respects, and provides a solution of the problem which, from the experimental point of view, leaves nothing to be desired.

3. The Vacuum Calorimeter † : First Form.—The principle of the method underlying this apparatus is simply that the substance under examination itself serves as calorimeter and is warmed a few degrees (or even less) by a platinum

---

* As A. Magnus has shown ("Physik. Zeitschr.," 1913, p. 5), the copper calorimeter may also be used for the determination of specific heats at high temperatures, if a large block of copper is employed.

† See also Supplement, p. 264.

SPECIFIC HEATS OF SOLIDS

wire into which is passed a measured quantity of electrical energy: the rise in temperature is also determined by means of the platinum wire, which functions at the same time as a resistance thermometer.

On account of the influence of the surroundings there would be, however, in normal circumstances, a temperature gradient inside the calorimeter (except where metals of high conductivity are concerned), which would prevent the possibility of any accuracy of measurement. To obviate this the calorimeter is placed in a vacuum as perfect as possible. Since at low temperatures the effect of radiation becomes negligibly small, it is always possible to obtain such excellent thermal insulation that the above source of error becomes completely negligible. In certain cases it is possible to reduce practically to zero the interchange of heat between the calorimeter and its surroundings, but this is of course not essential for the carrying out of good measurements; it suffices, in short, if the times of cooling are large compared with the time required for obtaining uniformity of temperature in the calorimeter.

Gaede * has, it is true, used a similar method in work with metals, only without the employment of a vacuum; but for the latter, the experiments here described could not have been carried out.

In 1909 my then private assistant, Dr. A. Eucken (23), undertook, at my suggestion, the development of the above method, and he solved the problem so well that he was able to clear up all the essential points which cropped up.

After the method had thus been shown to be quite practicable, I undertook (47), with the capable co-operation of my private assistant, Dr. Pollitzer, the further development of the method, especially for the domain of the lowest temperatures; we carried out a large number of measurements, to the results of which we shall frequently have to refer.

* W. Gaede, "Physik. Zeitschr.," **4**, 105 (1902).

In Fig. 4, K is the calorimeter proper, which is hung from the two lead-in wires; it is contained in a pear-shaped vessel which is evacuated as completely as possible by means of a Gaede pump, and usually also by means of cocoanut charcoal which, having been previously strongly ignited * *in vacuo,* was cooled in liquid air.

The measurement of resistance was carried out by means of a calibrated Post Office bridge; the energy introduced was determined by means of accurate voltmeters and ammeters, the current flowing through the voltmeter being, of course, subtracted from the indication given by the ammeter.

As calorimeter vessels the three forms illustrated below were employed. Metals, on account of their high thermal conductivity, could be used without any enclosure simply in the form of a block; this was of cylindrical shape, and was provided with a hole in which was inserted a core made of the same metal and wound with platinum wire. Thin waxed paper was used for insulation; the very small space between the core and the block was filled up with melted paraffin wax. The upper part of the core was

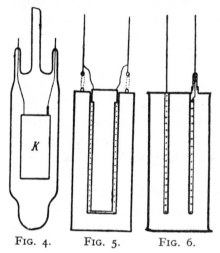

FIG. 4.   FIG. 5.   FIG. 6.

* The charcoal was contained in a tube of difficultly fusible potash glass, which was sealed by means of a glass of special composition to the remaining glass parts, and so could be heated to redness; the connection between the two kinds of glass is, however, more conveniently made by means of a ground joint with a mercury seal; even cementing with good white shellac will serve.

SPECIFIC HEATS OF SOLIDS

somewhat wider, so that it could be pushed into the hole to assist exchange of heat by a good contact (Fig. 5, *ca.* half natural size).

Non-metallic substances were introduced into the silver vessel shown in Fig. 6; this has the platinum winding on a silver tube soldered inside it, which also facilitates equalization of the heat. The winding was done as described above: insulation was secured either by means of shellac or varnish and tissue paper. The platinum wire has one end soldered to the silver vessel; the other end passes insulated through a little platinum tube soldered on to the silver: the end of the tube is sealed airtight with glass enamel. Further protection is given to the platinum winding by tin-foil wound round the inner tube. It is absolutely necessary that air should be present inside the silver vessel to facilitate equalization of temperature; consequently, after introducing the substance the lid at the bottom was carefully sealed on. If the silver vessel leaked, as it did occasionally, this was betrayed by a rapid increase in temperature of the platinum wire during the heating, and by the inordinately slow equalization of heat.

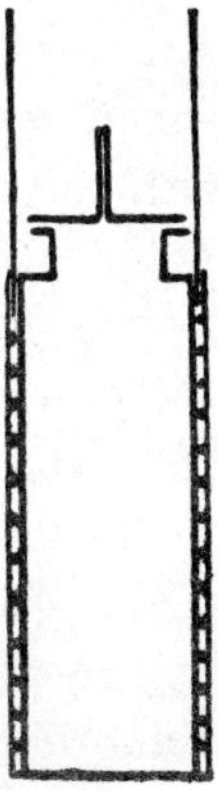

The third form (Fig. 7) differs only in detail from the second; it was preferred for the experiments with liquid hydrogen, because it could be made of much smaller dimensions. The platinum wire was wound on the outside of the cylindrical silver vessel and covered, to avoid thermal losses, with silver foil which was soldered at the edges to give a better thermal contact; this form has the advantage that the platinum wire does not have to be introduced vacuum-tight into the inside of the silver vessel. In a small size and at low temperatures this form of calorimeter proved to be excellent. The heat capacity of the silver vessel could be calculated with good accuracy, but it was also directly determined by a series of

FIG. 7.

measurements at different temperatures, the empty vessel in question being fused into the glass bulb.

The first two forms of calorimeter described were used down to the temperature of liquid air, of which the boiling-point could be lowered by about another 20° by reducing the pressure. These researches were mostly carried out during the winter 1909-1910. In the light of the experience so obtained it appeared attractive to continue the measurements down to the temperature of boiling hydrogen.

There was nothing to be altered in the experimental arrangement, for the method already described still gave, as was to be expected, any desired precision at these temperatures. The calibration of the platinum wire used occasioned some difficulties, but eventually we were able to fix with certainty a number of temperatures and to relate them by an equation, such that the very variable temperature coefficient for the region 20° to 40° abs., which was what was really wanted, could be derived with sufficient accuracy (compare below).

My first experiments were made with cylinder hydrogen, which was liquefied in small home-made apparatus; into the construction of this apparatus, which has been described in another place (57), I shall not here enter, remarking only that the idea of it was to introduce the evacuated glass vessel containing the substance under investigation straight into the liquefaction vessel; as this avoided the necessity for filling entirely with hydrogen, it was possible to work with only a few grammes of liquid hydrogen. The experiments with this apparatus run so simply and smoothly that one can carry out routine measurements without much trouble or expense. The nature of the method used by me is, of course, such that by gradual warming up from the lowest temperature the variation of the true specific heat can be determined forthwith up to much higher temperatures.

In order to be able to work with larger amounts of liquid hydrogen, which was also desirable for other purposes, a

SPECIFIC HEATS OF SOLIDS

compressor was installed of the type supplied with the smallest sizes of the well-known Linde liquid-air apparatus; this was kindly placed at my disposal by Herr Rubens. It compressed about $2\frac{1}{2}$ cubic metres of hydrogen per hour to 150 or 200 atmos.; the above-mentioned apparatus for hydrogen liquefaction liquefied about a tenth of this, and so yielded 300 to 400 c.c. of liquid hydrogen per hour. The great advantage of the compressor is that by means of it practically the whole content of an ordinary hydrogen cylinder may be liquefied, whereas if only the pressure of the cylinder hydrogen is available, only one-twentieth of it, at the most, can be liquefied and the remaining hydrogen must be allowed to go to waste.

After the compressor had been installed, it appeared more convenient simply to immerse the glass vessel, shown in Fig. 4, in a vacuum flask containing 200 to 300 c.c. of liquid hydrogen, just as when working with liquid air. Further, the vacuum, which proved to be critical for the experiments at the lowest temperatures, was improved by dipping a tube attached to the side and containing wood charcoal, as above, into a small vacuum vessel of liquid hydrogen.

In these experiments the silver vessel which contained the substance to be examined had, of course, to be filled with hydrogen instead of with air, in order to facilitate equalization of temperature. To this end there was a little silver capillary in the cover which was rapidly closed airtight with a drop of solder after filling with hydrogen. The glass vessel containing the calorimeter had also to be filled initially with hydrogen, at about 1 cm. pressure, which was pumped off after the temperature of the calorimeter had fallen sufficiently low.

In the experiments at the very lowest temperatures the vacuum had to be particularly good if it was desired to work with a very small "drift" of calorimeter temperature. This is because the pump used evacuates only to a certain pressure; hence the density of the gas, and consequently the number of molecules, which determines the thermal conductivity at low pressures, in the glass vessel containing the calorimeter is much greater than corresponds with the efficiency of the pump. Then again, the heat capacity of the calorimeter generally falls to a very small fraction of that at higher temperatures, and hence the calorimeter changes its temperature more rapidly, even though the conductivity of the surroundings is less. With careful working, the ordinary Gaede rotating mercury pump is good enough, but the highest accuracy can only be attained, at any rate at very low temperatures, when practically all the gas surrounding the calorimeter is removed, as above described, by means of charcoal cooled in liquid hydrogen.

In such a vacuum it is sufficient, when working with liquid hydrogen, if only a small amount of liquid remains in the bottom of the vacuum flask, once the calorimeter has cooled down, because the cold hydrogen vapour has a good thermal conductivity. Even after all the hydrogen has evaporated, the measurements can easily be continued at the higher and higher temperatures corresponding with the successive heatings of the calorimeter. By replacing the outer vessel of liquid hydrogen at appropriate times with baths of liquid air, carbon dioxide, and ice *seriatim*, it is thus possible to determine the true specific heats in a continuous series of experiments from $T = 15°$ (temperature of rapidly pumped off liquid hydrogen) up to room temperature.

4. Measurement of Temperature.—Under favourable conditions the accuracy of the measurements is practically dependent only on the accuracy with which the platinum wire used has been calibrated. It is not sufficient to calibrate a sample of the wire in question once for all, but some control determinations at least must be made for every calorimeter used, for the temperature coefficient of the wire is slightly influenced by the manner in which it is embedded. The following methods were tried :—

1. As fixed points, in addition to the temperature of melting ice, were used the boiling-point of carbonic acid snow,* of oxygen,† and of hydrogen ‡ : the two former substances were introduced into the excellent thermometer described by Stock § in the purest possible state, while in the last case it proved more convenient to allow the hydrogen contained in the vacuum flask to boil at different pressures. In this manner it was possible not only to determine the resistances at the fixed points themselves, but also to find with great accuracy, from the known vapour pressure curves of the substances mentioned, the temperature coefficient of the resistance in the neighbourhood of these fixed points.

2. In addition, a number of measurements were made with an air thermometer, the details of which will not, however, be here described.

3. Further, the platinum wire used was compared with a resistance made of the purest lead wire : in this connection the very regular curve given by lead, according to the measurements of Kamerlingh-Onnes,‖ was of great use, and it was also found that the pure lead used by me, which was obtained from Kahlbaum, showed the same resistance at the above fixed points, as had been found by Kamerlingh-Onnes.

4. Finally, a simple empirical rule (49) was discovered to connect the resistance curve of my platinum with that of the samples of platinum used by Kamerlingh-Onnes and Clay. For if the resistances of two samples of platinum are denoted by W_1 and W_2, and the resistance at $0°$ C. is taken as 1·000 as usual, then the equation,

$$W_1 = \frac{W_2 - a}{1 - a} \quad . \quad . \quad . \quad . \quad (8)$$

* L. Holborn, " Ann. d. Phys.," **6**, 245 (1901).

† H. Kamerlingh-Onnes, " Comm. Phys. Lab. Leiden," No. 107 (1908).

‡ M. W. Travers, " Experimental Study of Gases."

§ A. Stock, " Ber. d. deutsch. chem. Ges.," **39**, 2066 (1906) ; the form given by H. v. Siemens (91) is convenient.

‖ H. Kamerlingh-Onnes, " Comm." No. 99 (1907).

holds, where the figure a, independent of temperature, must be small compared with unity, as is of course always the case with moderately pure samples of platinum. Hence we have for the temperature coefficients

$$\frac{dW_1}{dT} = \frac{dW_2}{dT} \cdot \frac{1}{1-a}.$$

The above rule has been found to be of sufficient accuracy for freely supported or loosely wound platinum; it does not hold, however, for platinum fused into quartz (Heraeus' resistance thermometer), and it is subject to small inaccuracies when the wire is embedded in a solid varnish.

By the use of the above four methods, and also by combination of my calibrations with the numerous investigations of Kamerlingh-Onnes, I obtained the following table, the values in which were deduced, some by calculation and some by graphical interpolation: they are, of course, correct only for one particular sample of platinum.

TABLE II
Resistance Curve of Platinum

T.	W.	$\frac{dW}{dT} \times 10^3$.	T.	W.	$\frac{dW}{dT} \times 10^3$.
2	0·0091	(0·00)	90	0·2507	4·27
4	0·0091	(0·00)	100	0·2934	4·25
6	0·00915	(0·04)	110	0·3360	4·23
8	0·00925	(0·10)	120	0·3782	4·21
10	0·00955	(0·18)	130	0·4205	4·19
12	0·0100	(0·27)	140	0·4624	4·18
14	0·0106	0·39	150	0·5040	4·15
16	0·0114	0·52	160	0·5450	4·13
18	0·01255	0·67	170	0·5862	4·10
20	0·0140	0·85	180	0·6270	4·07
25	0·0195	1·28	190	0·6676	4·05
30	0·0270	1·91	200	0·7080	4·04
35	0·0365	2·40	210	0·7483	4·03
40	0·0490	2·90	220	0·7885	4·02
45	0·0648	3·38	230	0·8287	4·00
50	0·0834	3·72	240	0·8887	3·98
60	0·1235	4·02	250	0·9284	3·97
70	0·1650	4·20	260	0·9681	3·96
80	0·2080	4·28	270	0·9877	3·95
			273·1	1·0000	3·94

SPECIFIC HEATS OF SOLIDS

Quite recently Eucken and Schwers (84) have made some measurements with the vacuum calorimeter, which were planned particularly to test the T^3-law of Debye (cf. following chapter); in view of this object the highest accuracy was wanted at the lowest temperatures (in thermodynamical calculations the specific heats are of subordinate importance in this region because of their smallness). The method used was essentially that described above; the heating was done, however, with a constantan wire, while the measurements of temperature were made by means of a lead wire, the resistance curve of this material at very low temperatures being of a more suitable form than is the case with platinum. From the measurements of Kamerlingh-Onnes and Clay the following table was calculated by the above-mentioned authors, and was related to the sample of lead used by them by correction with the aid of the α-rule (p. 36).

TABLE III
Resistance Curve of Lead

T.	W.	$\frac{dW}{dT} \times 10^3$.	T.	W.	$\frac{dW}{dT} \times 10^3$.
14	0·01212	2·48	45	0·12638	3·95
16	0·01733	2·73	50	0·14598	3·90
18	0·02305	2·98	55	0·16537	3·86
20	0·02928	3·24	60	0·18458	3·82
22	0·03603	3·40	65	0·20354	3·78
24	0·04325	3·71	70	0·22236	3·75
26	0·05082	3·85	75	0·24105	3·73
28	0·05862	3·93	80	0·25968	3·72
30	0·06654	3·97	85	0·27825	3·71
35	0·08652	4·01	90	0·29676	3·70
40	0·10653	3·99	273·1	1·0000	—

It is obvious that interpolation of $\frac{dW}{dT}$ can be done with much greater safety for lead than for platinum at low temperatures.

5. The Vacuum-Calorimeter: Second Form.—In the

first form the calorimeter was electrically heated by means of a platinum wire, which served also as a resistance thermometer. The accuracy of the measurements depends (a) on the accuracy with which the change in resistance produced by a known input of electrical energy, corrected if necessary for thermal losses, can be determined; and (b) on the accuracy with which the temperature coefficient of the platinum wire used is known in the range of temperature concerned.

Now, as has already been remarked, this temperature coefficient varies at low temperatures in a manner not very favourable to accurate measurements; there is a fairly sharp maximum in the neighbourhood of 80° abs., and it falls very rapidly below 40° abs. In spite of the most careful calibration, which presents difficulties in the region from 25° to 45° abs. on account of the lack of fixed points, there are still uncertainties of a small amount. There is a further inconvenience in the fact that the platinum wire has its resistance considerably changed by the introduction of the electrical energy, and that, in consequence, the accurate determination of the latter, though quite possible, is nevertheless somewhat troublesome.

These disadvantages may all be avoided, and are in fact considerably reduced by the use of lead instead of platinum wire, but a simpler procedure is still desirable. The problem of establishing the variation of specific heat for a large number of substances as completely as possible is extremely important, both in the light of any future theory of the solid state, which will apparently be developed from a consideration of low-temperature properties, and also in view of the applications of our Heat Theorem and the study of chemical affinity.

Little had been achieved in this connection when I set myself the task of examining a few substances with the greatest possible accuracy; but the arbitrariness which would have existed in their selection would have weighed

SPECIFIC HEATS OF SOLIDS

against any advance of theory. It was only by the measurement of numerous examples of the most different classes of substance that it was possible to decide with certainty on the simplest type, which was then the occasion of a theoretical examination (cf. Chapter IV). Since there is a large mass of material awaiting further working out, any simplification of the methods already described is a distinct gain.

Such a simplification may be achieved, at any rate in certain respects, by altering the method of measurement of temperature. In the second form of vacuum-calorimeter now to be described, which I developed in collaboration with Dr. F. Schwers (95), the variation of temperature resulting from the electrical heating was therefore measured by means of a thermo-couple. One junction was soldered to the calorimeter, while the other lay on a block of lead, the temperature of which remained very nearly constant during the measurement; the small changes which occurred took place so slowly and uniformly that they were completely eliminated when applying the correction for the drift in temperature of the calorimeter. In order to know the actual temperature of the calorimeter, that of the leaden block had of course to be determined from time to time. The heating was done by means of a constantan wire put inside the calorimeter; since this wire has a temperature coefficient which is negligible even at the lowest temperature used, the current and voltage remain quite constant during the heating, a circumstance which is conducive equally to convenience and to accuracy.

To the above-mentioned lead block was soldered a copper sheath which surrounded the calorimeter; the effect of this was that the latter was in a space having an absolutely uniform external temperature. This protection might have been supposed superfluous, as neither radiation nor conduction is appreciable when the vacuum is good and the temperature low; but it was found that if it were omitted the deflections of the galvanometer connected to

40 THE NEW HEAT THEOREM

the thermo-couple had not by any means the same regular course as with it. From the experimental point of view, the copper sheath must, therefore, be regarded as a decided improvement.

The arrangement of the apparatus may be seen from

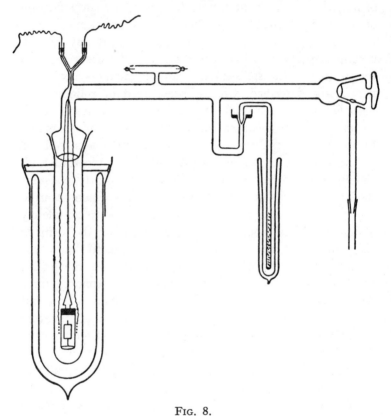

Fig. 8.

Fig. 8. A large Dewar flask, which can be closed by a rubber cap and cork, serves as a cooling bath; the stopper renders it possible to pump out the hydrogen from it, and thus to lower the temperature still further. A long cylindrical vessel serves to contain the calorimeter and its sheath: the opening in the top is closed by a glass stopper of suitable

SPECIFIC HEATS OF SOLIDS

width. Through the latter are introduced the necessary wires (in our case 8), this being done by taking them through a thin glass tube which was very carefully sealed with white shellac and picein. There was also a very wide tube leading from the glass stopper to a Gaede molecular air pump; at the end of this tube was a stop-cock, also of very wide bore, so as to utilize to the utmost the action of the pump.

The vacuum given by this excellent pump * was found, however, to be insufficient even when it was run continuously. This is explained in part by the known fact that hydrogen is not very well exhausted by this pump; hydrogen was present in the apparatus of necessity, having been used to cool the calorimeter before the experiment proper. Washing out with helium was of some benefit, and neon would probably have worked even better—other gases are, of course, of no use at temperatures of 15° to 20° abs. on account of their small vapour pressure.

It was therefore necessary to complete the evacuation by the use of ignited cocoanut charcoal, which was cooled with liquid hydrogen. In order to make the charcoal sufficiently active, a previous strong ignition *in vacuo* is desirable, and it was therefore contained, as in the earlier experiments, in a tube of difficultly fusible glass.

The Geissler tube indicated in Fig. 8 served to test the vacuum; but it is not suggested that, when the heat capacity of the calorimeter is small, the vacuum is good enough if no discharge passes through the Geissler tube. The experiment itself affords the real test of the vacuum; the drift of the temperature of the calorimeter, after it has been electrically heated, must be satisfactorily slow. When carbon and liquid hydrogen were used the vacuum was always excellent.

* Apart from the saving in time, the Gaede rotating mercury pump would have been equally satisfactory, perhaps even preferable, as it would have avoided the vigorous vibration which was frequently objectionable when working with the molecular air pump.

The arrangement of the calorimeter itself is illustrated in Fig. 9. The lead block, shown shaded, is seen above with the copper sheath fitting on to it. The sheath is wound about its middle with thin lead wire, from each end of which two connections are taken, so that the temperature of the sheath may be measured. The lead wire was insulated by winding it on thin paper, and over it was stuck paper and, finally, tin-foil. The resistance was measured by the potentiometric method in which, of course, the influence of the connecting wires is completely eliminated. A direct-reading galvanometer was attached, first, to the end of the lead wire and, second, to the ends of a known resistance in series with the lead wire; by comparing the deflections the resistance of the wire could be very rapidly determined at any moment to 1-2 parts per thousand, and the temperature of the copper sheath, therefore, to better than 0·1°, which was more than sufficient.

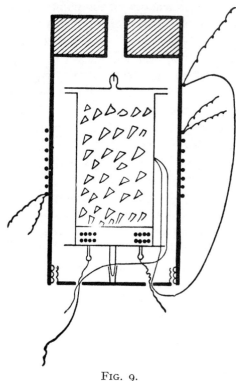

FIG. 9.

Much experience has been accumulated in our Institute on the behaviour of such lead thermometers [cf., e.g. Bulle (85)]. It is necessary to determine the resistance of

SPECIFIC HEATS OF SOLIDS

the lead wire frequently at the temperature of melting ice, for it slowly increases in the course of time. By adopting this precaution any marked inaccuracy may be easily avoided. The temperatures were ascertained from the resistance with the aid of the table calculated by Eucken and Schwers (p. 37); the α-value of the lead winding used was fixed by determining the resistance in boiling hydrogen; Schimank (94) has recently demonstrated that this rule is capable of giving as good service with lead wires as with platinum.

The calorimeter was made of copper, instead of silver as before, because the decrease of specific heat is then more rapid. As the dimensions were fairly small (cf. Fig. 9), its water value, which was carefully determined by special experiment, was frequently an appreciable fraction of the heat capacity of the calorimeter full of any particular substance. An increase of accuracy would have required the use of a larger calorimeter and a concomitantly larger sheath, but then the consumption of liquid hydrogen would have been correspondingly greater. On the lower lid of the calorimeter two thin tubes of platinum were soldered, through which were led the insulated platinum wires. An hermetical seal was made with enamel glass. Between these platinum wires, and practically on the bottom of the vessel, lay the spirally wound constantan wire, of which the resistance amounted to nearly 100 ohms. On this wire rested some thin unvarnished tissue paper, which is quite permeable to hydrogen, and over it the substance under investigation was piled. The upper lid was soldered on and unsoldered to get the substance in and out, and care was taken always to use the same amount of solder. In the middle of the upper lid was a thin copper tube which, after pumping out and filling with hydrogen, was closed by a well-fitting peg, and then soldered. With the arrangement described equalization of temperature inside the calorimeter, after heating was finished, required only about a minute (compare the example calculated below). The calorimeter was supported, as shown, on a thin ebonite rod. The attachment of the two junctions of the thermo-couple is also evident from the figure.

I tried at first to find a thermo-couple which would give readings proportional to the temperature difference in the interval from 15° to 100° abs. ; as the detailed investigation of Wietzel (86) shows, theprobability of finding such a couple is not great. Eventually, therefore, a copper-constantan couple was used ; the direct connection between the two junctions (cf. Fig. 9) was made with fine constantan, the two leads being of copper.

Calibration of the couple used was done by two different methods. A fairly long piece of homogeneous fine constantan wire was selected by the usual methods, and the thermal E.M.F. measured as accurately as possible, when one junction was in melting ice and the other in a series of baths at different low temperatures. These temperatures were determined by means of the Stock's thermometer calibrated by H. v. Siemens (91). For the lowest temperatures hydrogen was used, boiling either under the ordinary pressure or under an accurately measured reduced pressure. The values could be fairly well reproduced by the following interpolation formula,

$$E = 31\cdot32T \log_{10}\left(1 + \frac{T}{90}\right) + 1\cdot0 \times 10^{-7}T^4 \text{ microvolts,}$$

where E is the electro-motive force between the absolute zero and the temperature of experiment.

With the aid of the above formula, values of $\frac{dE}{dT}$ could be calculated for every degree ; a table was thus obtained which came very near to the truth.

As a further control, the copper calorimeter of Fig. 9 was replaced by the lead block accurately examined by Eucken and Schwers (p. 37). The above table was corrected with the aid of these numbers ; this could easily be done as the

SPECIFIC HEATS OF SOLIDS

corrections were found to be negligibly small at the higher temperatures, and amounted only to a few per cent. at the lower. In the neighbourhood of room temperature, where the above formula becomes inaccurate, $\frac{dE}{dT}$ was found independently by a second series of experiments, a matter which naturally presented no difficulties.

The following table contains an extract from the numbers so deduced. It shows the great simplification resulting from the fact that $\frac{dE}{dT}$ varies much more regularly than does the temperature coefficient of platinum.

TABLE IV
Copper-Constantan Thermo-couple

T.	$\frac{dE}{dT}$	T.	$\frac{dE}{dT}$	T.	$\frac{dE}{dT}$
15·5	4·08	60·5	12·68	150·5	23·26
20·5	5·26	70·5	14·10	170·5	25·20
25·5	6·42	80·5	15·43	190·5	27·30
30·5	7·50	90·5	16·63	210·5	29·5
35·5	8·47	100·5	17·70	230·5	31·3
40·5	9·37	110·5	18·80	250·5	33·1
45·5	10·24	120·5	19·90	270·5	34·9
50·5	11·10	130·5	21·06	290·5	36·5

It may be seen that the fall in $\frac{dE}{dT}$ at low temperatures, though uniform, is very considerable. During the measurements it was soon found that this at first unwelcome behaviour offered great advantages. The heat capacity of the calorimeter, either with or without its filling, also falls rapidly with the temperature, and the law of its decrease is usually very similar to that of the decrease of $\frac{dE}{dT}$. When both curves run parallel, as was the case in some instances for a considerable range of temperature, the deflection of the galvanometer divided by the energy introduced is independent of the temperature. In this case the variation of the heat capacity is therefore given by the very accurately drawn calibration curve of the above table; in other cases the ratio varies much less with temperature than it would have done had $\frac{d\mathrm{E}}{d\mathrm{T}}$ been independent of temperature; this circumstance is of course favourable to the measurements.

The above numbers agree very well, moreover, with the values of Wietzel (86), who used a different sample of constantan, except that Wietzel's values are consistently higher by a constant 9·8 per cent.; it would be a great simplification in the use of such thermo-couples at low temperatures if this behaviour were common to all samples of constantan.

By way of further explanation, the numerical values obtained in one experiment may be briefly given. We shall select an example, haphazard, in which the calorimeter contained 38·31 gm. = 1·414 mols of aluminium. The temperature of the lead block was 21·0° abs. Heating was done for 40·2″ with 2·02 volts and 0·0202 amps.; these values were read on carefully calibrated Weston instruments. The quantity of heat amounted to

$$2·02 \times 0·0202 \times 40·2 \times 0·2388 = 0·392 \text{ gm.-cals.}$$

On account of the resistance of the leads to the constantan wire and the fall in potential across the ammeter, the energy actually introduced into the calorimeter is only 0·387 gm.-cals. The E.M.F. of the thermo-couple was measured with the aid of a small Siemens' mirror galvanometer by Dieselhorst: without this excellent instrument, which comes to rest in a few seconds, and is of ample sensitivity, the investigation would have been much more difficult. In the galvanometer circuit (in addition to a 300 ohm resistance) was a standard ohm to which a Weston cell with 100,000 ohms in series could be connected for occasional determination of the sensitivity. On reversing the connection to the cell in this series of experiments a difference

SPECIFIC HEATS OF SOLIDS

in reading of 30·0 mm. was obtained. The deflections of the galvanometer were read off with a glass scale and telescope, so that 0·1 mm. could be safely estimated.

The changes in the readings were as follows :—

Time.	Reading.
19′	185·8
20′	185·7
21′	185·8
Heat applied from 21′ 10″ to 21′ 50″.	
22′	201·8
23′	200·0
24′	199·0
25′	198·0
26′	197·0
27′	196·1

From 23′ onwards the drift is quite regular ; the equalization of temperature in the calorimeter was therefore complete in about a minute. By extrapolating the galvanometer readings to 21′ 30″ from both directions as usual, the heating effect is readily calculated as

$$201·6 - 185·8 = 15·8.$$

The galvanometer zero was at 172, as was found by reversing the leads from the thermo-couple by means of a mercury commutator free from thermal effects ; the initial temperature of the calorimeter is hence calculated as $21·0 + 1·7 = 22·7°$. The above heating effect corresponds with a rise in the temperature of the calorimeter of

$$\frac{15·8}{14·7 \times 0·602} = 1·786°,$$

and the heat capacity of the calorimeter plus contents is calculated as

$$\frac{0·387}{1·786} = 0·217 \text{ gm.-cals.},$$

valid for the mean temperature of $22·7 + \frac{1·8}{2} = 23·6°$. A great number of further measurements could be immediately proceeded with. After the deflection of the galvanometer from its zero had increased to 60 mm., the lead wire on the copper sheath was heated by a current of suitable strength; the temperature difference between the sheath and the calorimeter was thus again equalized, whereupon fresh measurements were at once undertaken.

After the temperature of the calorimeter had risen sufficiently, in virtue of the successive heatings, the large vacuum vessel containing liquid hydrogen, which is shown in Fig. 8, was replaced by another of liquid air. The small vacuum flask, also filled with hydrogen, was removed for a minute or two in order to equalize rapidly the temperature of the calorimeter and vacuum sheath; by again cooling the charcoal with liquid hydrogen a perfect vacuum was then restored. A series of measurements at higher temperatures could then follow.

As soon as a good vacuum is obtained the temperature of the copper sheath is always rising, owing to conduction of heat along the connecting wires; it is therefore necessary to have these wires as thin as possible.

One should not neglect, before constructing the apparatus, to obtain an idea of the heat balance of the calorimeter by calculating the heat conducted along the connecting wires; when carbon and liquid hydrogen are used to obtain the vacuum, this conduction plays the chief part, and consequently one attempts to make it as small as circumstances will allow.

The vacuum calorimeter described by Schwers and me has been further developed recently in a few respects by Gunther (107), who used it for many important measurements; reference may therefore be made to the description given by Gunther for a more detailed account. The vacuum calorimeter has been used by Eucken (105) with marked success to determine the specific heats of gases at constant volume, and also for the measurement of heats of evaporation, heats of fusion, and heats of transformation at low temperatures (cf. also Chapter V).

SPECIFIC HEATS OF SOLIDS

As regards comparison of the two forms of the vacuum calorimeter described, the former is capable of greater accuracy while the latter has the advantage of extreme convenience and rapidity of manipulation with an accuracy which is usually ample. The one or the other form will therefore be preferred according to circumstances.

6. Critical.—Upon the measurements which have been carried out by the methods described in this chapter depend, in part, the recent theories of the solid state (cf. Chapter IV), and also to a large extent the exact tests of our Heat Theorem. It will therefore be of use to devote a few words to the reliability of the results obtained.

From the beginning it was clear to me that it would be of little significance for our purpose to investigate a few substances with the highest accuracy. It was undoubtedly of much more importance to secure, at first in its general lines, a picture as complete as possible of the behaviour of the specific heats at low temperature, avoiding, of course, as far as possible serious errors of observation.

To this end I searched for methods which should be very different, and therefore mutually confirmatory; I think I have found them in the above described copper calorimeter and vacuum calorimeter. The latter apparatus is, of course, much more suited to our problem, since it gives the true specific heats (or, rather, the mean specific heats for a very small interval of temperature); but the copper calorimeter has also rendered us valuable service in completing, and above all in controlling, the results. For wherever a mutual control was possible—and the number of such examples is very large—an extremely satisfactory concordance has been shown between the values obtained; this, of course, says much for the reliability of both methods.

Any measurement of specific heat depends on the accuracy of the temperature scale employed; it was not quite simple, particularly at the time when I began my measurements in the neighbourhood of the boiling-point of hydrogen, to secure adequate reliability in this respect. In view of the different mutually complementary methods described on pages 35 *et seq.*, this object seems to have been attained with an accuracy sufficient for my purpose.

It is necessary for me to deal here with a remark of Kamerlingh-Onnes ("Communications," No. 143, p. 5, 1914), which might easily give rise to an incorrect idea on the above question. For, according to him, the temperature scale used by me, and later by Eucken, is based " mainly " on the older calibration of Pt_1 by Kamerlingh-Onnes, Brook, and Clay, which should be replaced by the more recent determinations of Kamerlingh-Onnes and Holst.

Now, I should certainly never have undertaken a series of investigations so extensive as my work on the specific heat of solids was bound to be, without having some other basis than the calibration of a particular sample of platinum, however eminent the observer to whom it were due. I recognized, of course, the possibility of an erroneous determination by one of my predecessors, and consequently, as described in detail in my paper (47), and also on pages 34 *et seq.*, I made use of a series of different and mutually controlling bases, in particular the measurements of Travers on the boiling-point of hydrogen at different pressures (cf. also p. 35), confirmed later by Kamerlingh-Onnes, and also the lead curve of the latter. There can, therefore, be no possibility of small inaccuracies in the Pt_1-curve having appreciably affected my results.

In any case, it is not strictly true that I relied mainly on Pt_1 among the different samples of platinum; the α-rule, though only an approximation formula, holds so well for other samples of platinum (cf. 49) that its use excluded any decided preference for a particular sample.

But fortunately we have also quite a direct method of testing the question as to whether there does exist an appreciable difference between the temperature scales used by Kamerlingh-Onnes, on the one hand, and by me and my co-workers on the other.

SPECIFIC HEATS OF SOLIDS

Kamerlingh-Onnes and Keesom * have quite recently measured, by a method essentially similar to that described in Section 3, the atomic heats of lead, copper, and of solid and liquid nitrogen.

Lead, in particular, was investigated with special care as a suitable standard substance (cf. p. 43) by Eucken and Schwers (84) : these are probably the most accurate measurements available in this region. The authors find that in the range from $T = 16°$ to $80°$ their measurements are represented almost perfectly by Debye's function with the value $\beta\nu = 88$. Kamerlingh-Onnes finds that in the same range of temperature complete agreement is again obtained with $\beta\nu = 88$, both at the lower and at the higher temperatures; in the middle of the range there are small deviations which he thinks slightly exceed the errors of observation.† The agreement between the two series of numbers is at any rate as good as could be expected.

Further, my own older measurements (47) agree well with the more recent values of Kamerlingh-Onnes and Keesom :—

		T = 23.0	28.3	36.8	38.1	85.5
Nernst	. $C_p =$	2.96	3.92	4.40	4.45	5.65
Onnes	. $C_p =$	3.06	3.65	4.48	4.52	5.70
Nernst (calc.)	. $C_p =$	2.96	3.64	4.37	4.45	5.68

The values of Kamerlingh-Onnes and Keesom are interpolated from their measurements, the last slightly extrapolated. My measurements with lead were among the first which I extended to very low temperatures. I had not expected, and in view of the newness of the subject could hardly have attempted, a greater accuracy than is shown by my values compared with those of Kamerlingh-Onnes.

---

* "Comm." No. 143 (1914), and " Kgl. Akad. d. Wissensch.," Amsterdam, 29th Jan., 1916.

† In view of its derivation, Debye's formula can only hold strictly in the T^3-region, and in the range of the Dulong and Petit law.

In the last row are values calculated with the aid of the formula of Lindemann and myself (61), and given as the most probable values at that time. Here the agreement with Onnes may well be regarded as ample for most purposes of thermodynamical calculation.

The accuracy of the figures in the middle horizontal line is further demonstrated by their close agreement, which has been mentioned above, with the measurements of Eucken and Schwers.

Copper was measured by Kamerlingh-Onnes over the range of temperature $T = 15°$ to $T = 22°$; Debye's formula, which owing to the low atomic heat of copper in this region reduces to the T^3-formula (19) of Chapter IV, agrees perfectly here with $\beta\nu = 323$, whereas I had derived $\beta\nu = 315$ (cf. 78) for higher temperatures ($T = 23°$ to $T = 450°$); the concordance between these values may again be regarded as sufficient for most purposes (cf. Chapter XVI).

The measurements of the atomic heat of solid and liquid nitrogen, which were published almost simultaneously by Kamerlingh-Onnes and Keesom and by Eucken (105), agree in some parts perfectly, in others, to about 2 per cent.

In these circumstances there can be no question of any marked discrepancy between the temperature scales used by Kamerlingh-Onnes, on the one hand, and my co-workers and me on the other. It is, of course, pleasing to have established the fact that our methods have also proved perfectly trustworthy in the hands of Kamerlingh-Onnes and Keesom.

It is obvious that attention must be paid to securing the highest purity in the substances investigated; but the physical condition may also exert an influence in certain circumstances.

Thus Schwers and I (95), repeating the measurements previously made by me (47) with silver iodide with a preparation of similar purity but crystalline, found values for the molecular heat appreciably (*ca.* 10 per cent.) smaller at the lowest temperatures, because the material originally used by me, though perhaps partly crystalline,* was largely amorphous: my previous measurements on silver iodide are thus not above reproach in that they related to a material not quite fully characterized as to physical state. Actually, amorphous substances have practically always higher specific heats, though there are exceptions.

There may be grave danger from a similar source of error, namely, allotropic change, which sometimes, strange to say, takes place fairly rapidly even at the lowest temperatures. Diatomic elements, such as O_2 and N_2, in particular show such changes, as Eucken proved in his measurements (105); careful attention must be paid to them if serious errors are to be avoided. When the changes take place quickly enough, as in the above cases, it is possible to make accurate determinations of the heats of transformation. As Gunther (107) found, iodine also exhibits a similar change at low temperatures; this caused the first measurements I carried out in the region of the lowest temperatures to be considerably too high. If crystals of iodine are exposed for about half an hour to the temperature of liquid hydrogen they become brittle and powdery in parts; a behaviour of this sort will sometimes serve to direct attention to disturbing phenomena of transformation, and, in any case, the contents of the vacuum calorimeter may easily be examined for such changes after a series of experiments.

* Cf. on this point Braune and Koref (98b).

CHAPTER IV

THE LAW OF DULONG AND PETIT

1. General Results from Our Measurements of Specific Heat.—In dealing with the form which the law of Dulong and Petit has assumed in the light of recent theory, we shall frequently be concerned with considerations of molecular theory, which are foreign to our particular subject. Nevertheless, it will be necessary for an understanding of our Heat Theorem to deal briefly with recent views on the subject of specific heats; the results obtained with the vacuum calorimeter described in the previous chapter have not been without influence on their development.

At the present time it may be put forward as a view which is experimentally justified, and is supported by the recent theoretical developments of physics, that *the specific heats both of crystalline and of amorphous solids assume negligibly small values at very low temperatures.*

The idea that the atomic heats of all solid, i.e. nongaseous, substances without exception become very small at very low temperatures, is first found developed in the paper presented by me on the 23rd December, 1905, to the "Göttinger Gesellschaft der Wissenschaften," which also contains the first presentation of my Heat Theorem. I there arrived at the result, directly contradictory to all earlier ideas, that the heat of evaporation at low temperatures always rises first, reaches a maximum, and then, and not until then, exhibits the usual decrease; according to the First Law this means that the specific heat falls off rapidly at very low temperatures. The equation

$$\lim \frac{d\mathrm{U}}{d\mathrm{T}} = 0 \text{ (for } \mathrm{T} = 0) \quad . \quad . \quad . \quad (9)$$

LAW OF DULONG AND PETIT

demands, further, that at low temperatures the above-mentioned small limiting value for the atomic heat of any element should be independent of the nature of the condensate.

In the Silliman lecture (1906) I next showed that this limit must be between 0 and 2, and I made calculations, using the value 1·5, which, as I pointed out, was selected only preliminarily.

Next appeared the paper of Einstein (*vide infra*) in 1907,

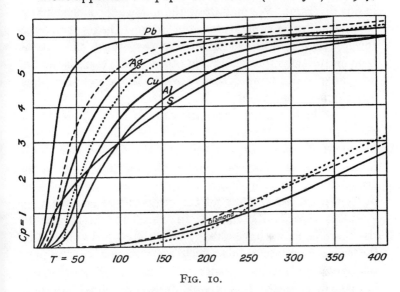

FIG. 10.

wherein Planck's formula for the energy of a resonator was applied to the atoms of solids; according to Einstein's formula all atomic heats at the lowest temperature must become not only zero, but zero of an infinitely high order.

This was where my measurements with the vacuum calorimeter were made, throwing new light on the above question from the experimental side. As the subject of the specific heats of solids at low temperatures is of fundamental importance for the purposes of this book, I may briefly illustrate in the accompanying diagram (Fig. 10) some of the most important results of these measurements.

The curves drawn full are the direct result of my measurements, and we notice at once that the curves for lead, silver, copper, and aluminium form a perfectly concordant family, in that they can be brought into coincidence by changing the scale of the temperature axis, i.e. by altering the value of a degree of temperature. The same is true for diamond, of which, however, only the bottom section can be reproduced in our diagram. We may draw the following conclusions:

1. At low temperatures the atomic heats converge towards zero in such a manner that they touch the axis of temperature asymptotically; in other words, there is always a point of finite temperature at which the atomic heat has a negligibly small value.

2. At high temperatures the atomic heats, reduced to constant volume, converge towards the value 3R required by the Law of Dulong and Petit; this is the case both for elements and for compounds (cf. also p. 57); in some cases, chiefly where hydrogen compounds are concerned, the melting-point lies too low for this region to be reached.

3. As regards the manner in which the atomic heats rise there is, as we have already seen above, a simple law in the case of certain solids which are apparently of a particularly simple constitution; this law is typified by the full curves, extrapolation of which to the absolute zero may be made with complete safety, taking into consideration the curve for diamond. Some simple chemical compounds crystallizing in the regular system, such as potassium chloride and sodium chloride, also follow the same law.

4. The law for this rise clearly recalls Einstein's equation for atomic heats, but the subsequent fall is considerably less steep; two examples of Einstein's family of theoretical curves, reduced to constant pressure, are shown in the dotted lines, and it is obvious that the experimental and theoretical curves intersect.

5. Practically all chemical compounds which have hitherto been examined show another and less steep decrease in their (mean) atomic heats. A few elements also behave similarly, e.g. sulphur: the curve for this element is reproduced in the above diagram.

These were the most important conclusions from my measurements; we now treat briefly of the theoretical developments associated with them, at least in so far as concerns the modification of Einstein's theory shown to be necessary by my experiments.

2. Theoretical.—As was shown by Boltzmann in 1871, statistical mechanics leads to the result that the mean kinetic energy of the atoms of solids must be as large as that of the atoms of a monatomic gas. Further, as long as the internal forces acting on any atom are proportional to its distance from the equilibrium position, or, more generally, are linear functions of the changes of its co-ordinates, the potential energy must be as large as the kinetic.*

Now however complicated the law of force between the atoms † may be, the force driving the atom back to its equilibrium position must, for sufficiently small displacements from rest, be proportional to its distance from that point; all theories of elasticity, for example, operate with this assumption. Thence it follows that at sufficiently low temperatures the Law of Dulong and Petit, or its generalization by Kopp and Neumann, must hold not only for elements but also for compounds.

The equations for energy content E and for atomic heat at constant volume C_v, are thus

$$E = 3RT, \quad C_v = 3R \qquad . \qquad . \quad \textbf{(10)}$$

Recent investigations have shown that the equation

$$C_v = 3R = 5 \cdot 96 \qquad . \qquad . \qquad . \quad \textbf{(11)}$$

* Cf. here Boltzmann, "Gastheorie," ii, p. 126; also "Theor. Chem.," p. 270.

† The law of force must, it is true, be a continuous one; according to Bohr's ideas, however, this is doubtful.

holds with great accuracy for many substances not too near the melting-point. In the neighbourhood of the melting-point higher values are mostly obtained; it is expressly to be noted that this happens both with substances which conduct metallically and with non-metals.* In the neighbourhood of the melting-point lead has $C_p = 6·61$, iodine $C_p = 6·64$, and the correction to C_v which has to be made by a well-known application of the Second Law (cf. p. 65) does not appear to account entirely for the excess above 3R.

There is no cogent reason for attempting to explain this excess above the normal value by means of any new hypothesis; it is reasonable to suppose that the above-mentioned conditions concerning the forces driving atoms back into their equilibrium position are no longer fulfilled.

The older theory can offer no explanation for the other much more important deviation from equation **(11)** which occurs without exception at low temperatures. It is at the lowest temperatures, when the oscillations about the mean position are the smallest, and should consequently best agree with theory, that the atomic heats of compounds fall to very low values.

Einstein † succeeded in giving a qualitative explanation of this behaviour by applying Planck's quantum theory to the oscillations of atoms about their equilibrium position. If this hypothesis be applied to each individual atom, and it be assumed that if ν is the oscillation frequency of the atom, the latter can only take up the quantum $\epsilon = h\nu$, or a whole multiple of it, we obtain

* Experiment has not up to the present given any support to the idea that the conducting electrons may have an influence on the specific heat, which should manifest itself at high temperature. Such an influence is, however, not of itself improbable, and possibly it may be experimentally demonstrated some day.

† "Ann. d. Physik.," (4), **22,** 184 (1907); cf. also "Theor. Chem.," pp. 272 *et seq.*

LAW OF DULONG AND PETIT

$$E = 3R\frac{\beta\nu}{e^{\frac{\beta\nu}{T}} - 1}, \quad C_v = 3R\frac{e^{\frac{\beta\nu}{T}}\left(\frac{\beta\nu}{T}\right)^2}{\left(e^{\frac{\beta\nu}{T}} - 1\right)^2}, \quad \beta\nu = \frac{h\nu N}{R} \quad (12)$$

(N = number of atoms per gramme atom.)

These formulæ failed from the quantitative point of view, but a modification found empirically by Lindemann and me (61) agreed very well with the available observations:

$$E = \frac{3R}{2}\left(\frac{\beta\nu}{e^{\frac{\beta\nu}{T}} - 1} + \frac{\frac{\beta\nu}{2}}{e^{\frac{\beta\nu}{T}} - 1}\right) \quad . \quad . \quad (13)$$

The whole subject reached a new stage with the investigations of Born and von Karman,* and in particular of Debye.†

Debye regards a solid as a *continuum*, and takes no notice of its atomistic structure. In accordance with a rule derived from the radiation theory for solids, he takes the number of oscillations in a solid body as

$$dz = A\nu^2 d\nu.$$

Now on the old theory, i.e. classical mechanics and the law of equipartition of energy, which latter holds in general as an exact consequence of classical mechanics, every oscillation must have the energy content $\frac{RT}{N}$, and the law of Dulong and Petit must also be obeyed; the old theory thus gives

$$E = 3RT = \frac{RTA}{N}\int_0^\nu \nu^2 d\nu = \frac{RT}{N}A\frac{\nu^3}{3} \quad . \quad . \quad (14)$$

---

* "Physikal. Zeitschr.," **13**, 279 (1912); **14**, 65 (1913).
† "Ann. d. Physik.," **39**, 789 (1912).

and it therefore follows that

$$A = \frac{9N}{\nu^3} \quad . \quad . \quad . \quad . \quad (15)$$

These equations hold, however, only for high temperatures; in order to bring the thermal motion into accord with Planck's radiation formula, we must, exactly as above, write for the energy content, instead of 3RT,

$$3R\frac{\beta\nu}{e^{\frac{\beta\nu}{T}} - 1};$$

consequently

$$E = \frac{R}{N}A \int_0^\nu \nu^2 \frac{\beta\nu}{e^{\frac{\beta\nu}{T}} - 1} \cdot d\nu \quad . \quad . \quad (16)$$

or, introducing the value found above for A,

$$E = \frac{9R}{\nu^3} \int_0^\nu \nu^2 \frac{\beta\nu}{e^{\frac{\beta\nu}{T}} - 1} \cdot d\nu \quad . \quad . \quad (17)$$

The upper limit indicated for the integral in the above integrations signifies the highest oscillation frequency of which the solid is capable; from the atomic standpoint this value is obviously to be regarded as the natural frequency of the atom, or at least a magnitude commensurable therewith.

The expression last obtained may be easily transformed into

$$E = 9RT\left(\frac{T}{\beta\nu}\right)^3 \int_0^{\frac{\beta\nu}{T}} \frac{x^3 dx}{e^x - 1} \quad . \quad . \quad (18)$$

The numerical evaluation of E or of

$$C_v = \frac{dE}{dT}$$

LAW OF DULONG AND PETIT

offers no difficulties.* Debye's equation reproduces excellently the course of the curves found experimentally by me for crystals of simple structure; two examples of Debye's curves are shown in Fig. 10 by the broken lines.

3. Debye's T^3-Law.—Debye's equation differs only slightly in form from equation (**13**), except for very low temperatures; for these temperatures it furnishes the remarkable result, easily deduced from the above equations,

$$E = aT^4, \quad C_v = 4aT^3, \quad a = \tfrac{3}{5}\pi^4 \frac{R}{(\beta \nu)^3} \qquad . \text{ (18}a\text{)}$$

It was naturally of extreme interest to test the above result experimentally on suitable examples, for according to it the specific heats of all solids, whether amorphous or crystalline, elementary or compound, become small to the third degree of infinity at very low temperatures. My measurements on diamond gave some indication, but a more detailed test was made on a series of substances by Eucken and Schwers (84), and then a still more exhaustive one by Nernst and Schwers (95). Quite recently (cf. p. 52), Kamerlingh-Onnes and Keesom were able to verify the law accurately for copper. To-day the T^3-law may be regarded as having been proved to be true as a limiting law valid for all solids.

4. General Formula for the Representation of Specific Heats.— In concluding this discussion, we shall also deal briefly with the question, important in the sequel, as to how the variations of the mean atomic heats may be expressed for compounds (and also for elements which, like sulphur, behave similarly to compounds).

---

* For the functions of Einstein, Nernst, and Lindemann, cf. the tables in Pollitzer, " Nernstches Wärmetheorem " (Enke, Stuttgart, 1912); for that of Debye, see, in addition to the above-mentioned paper, Nernst and Lindemann, " Sitzungsber. d. preuss. Akad.," 12th Dec., 1912, and also the tables at the end of this book.

The following considerations from molecular theory give us a hint on the subject.*

It is immediately obvious that the conception of molecular weight requires a definition for the solid state essentially different from that for gases; probably no suitable definition will be possible until we have a precise knowledge of the energy at the absolute zero. But it is, at any rate, certain that crystals, in which no atom differs from another in its mode of attachment or of motion and in which there are no definite groupings of several atoms, may conveniently be treated as monatomic; the diamond is a particularly instructive example in this respect.

In compounds, certain groups of atoms may often be recognized from the Laue diagram as specially distinguished by their geometric arrangement (e.g. the radicals $-CO_3$, $-NO_3$); consequently there will be a difference between the nature of those forces, on the one hand, which bind the atoms in the respective groups together, and those, on the other, which unite these groups with other atoms or groups of atoms.

For the sake of shortness we shall speak of any such grouping as a "molecule," and we arrive at the following conceptions. We shall have to suppose that in compounds the cohesion of the molecules in the crystal results from forces which reach from molecule to molecule, and it is reasonable to suppose that the cohesion is produced by valence forces of a nature similar to that assumed for the so-called molecular compounds. The molecules will execute thermal movements analogous to those of atoms in monatomic crystals.

In addition, the atoms within the confines of the molecule will oscillate to and from one another; at temperatures sufficiently high to exclude the quantum theory, each atom has an energy of vibration $3RT$ (law of Kopp-Neumann); at lower temperatures the mean energy of vibration per

* Nernst, "Göttinger Vorträge," p. 79 (Teubner, Leipzig, 1914).

LAW OF DULONG AND PETIT

atom is smaller, so that the mean atomic heat falls below the value 3R.

On these considerations the energy of oscillation of the molecule will decrease according to the equation of Debye, which we may regard as thoroughly established by experiment.

The energy of vibration of the atoms within the molecule must be in accordance with Einstein's function (equation for a simple resonator), provided that the number of atoms in the molecule is not too large, and that joint oscillation of several atoms, as suggested on page 62, is hence not possible.

Now Debye's function decreases with temperature much more slowly than Einstein's; this has already been mentioned above, but may be further illustrated by the subjoined short table :—

$\frac{\beta\nu}{T}$	Debye.	Einstein.
0·0	5·955	5·955
1·0	5·670	5·48
5·0	2·198	1·02
10·0	0·451	0·027
20·0	0·058	0·000

Hence, if we deal with sufficiently low temperatures, the energy content must be conditioned solely by the molecular vibrations, and we may then apply the laws which we have learned for monatomic substances. Assume that we determine, by some means, the value of $\beta\nu$ for the whole molecule, and measure the average atomic heat C_v for the sufficiently low temperature T'; then $F\left(\frac{\beta\nu}{T'}\right) = nC_v$.

Here the left side connotes Debye's value for the atomic heat calculated for $\beta\nu$ and the temperature T', and the right side signifies the molecular heat of the n-atomic molecule (C_v = mean atomic heat). From this equation we can calculate n directly, i.e. we can determine of how many atoms the molecule of the crystal consists. Thus we arrive for the first time at a determination of molecular weight for crystalline substances, which is just as direct as is the use of Avrogadro's law for gases.

Of course, experience must decide whether the above views are valid without limitations. Since the molecules are probably very densely packed in the crystalline state, it may not be impossible, *a priori*, that there are also direct forces acting between one atom and another in a neighbouring molecule; this would naturally influence the state of vibration. But the idea expressed above seems to me so simple that I attribute a certain degree of probability to it; in fact, it would simplify matters considerably if compounds could be treated as monatomic substances at temperatures sufficiently low for the energy of atomic vibrations in the molecule to have disappeared.

However that may be, the above considerations suggest the expression of the molecular heats of compounds by means of a suitable combination of a Debye function with one or more Einstein functions; as was to be expected, many elements (e.g. sulphur) behave here like compounds. For our present purpose it does not matter whether the above considerations from molecular theory are right or not; it is sufficient to establish the fact that the method of calculation mentioned has proved itself convenient in practice; examples are given in the lecture cited on page 62.

In support of this method of calculation the two following points may be particularly adduced :—

1. The sum of the functions used gives at low temperatures the T^3-law of Debye; it is not only, therefore, in agreement with the surest theoretical basis which we have in this region, but it gives us a useful extrapolation to the absolute zero, which is required especially in applications of our Heat Theorem.

2. In those cases, naturally rare, where the forces between molecules are large in comparison with the forces between

the atoms in the molecules, there must exist a range of temperature in which the increase occurs according to the Einstein formula before that according to the Debye formula has become appreciable. This may lead to an accelerated increase (i.e. an increase according to higher powers of T) occurring in the range considered, after an initial increase of specific heat has taken place according to the T^3-law. Cases of this sort have actually been observed by Schwers and me (95) and by Gunther (107); they are, of course, incapable of explanation from the standpoint simply of Debye's theory. It is noteworthy that the examples hitherto obtained are compounds of very high melting-point, for which the one assumption, namely, that of strong mutual attraction between the molecules of the crystal, is certainly fulfilled.

5. Calculation of C_v from C_p.—In the equations derived on pages 59 and 60, the atomic heat was always calculated at constant volume, C_v, whereas all the measurements relate, on the contrary, to C_p. Thermodynamics supplies the well-known relation

$$C_p = C_v\left(1 + T\frac{9a^2 V}{KC_v}\right), \quad . \quad . \quad . \quad (19)$$

wherein a denotes the coefficient of thermal expansion, V the atomic volume, and K the compressibility. As the data requisite for the application of the foregoing equation are available in only a few cases, Lindemann and I (61) have obtained an approximation formula,

$$C_p = C_v + 0\cdot 0214\, C_p^2 \frac{T}{T_s}, \quad . \quad . \quad (20)$$

where T_s denotes the melting-point. It is often possible, especially when one is only concerned with obtaining a good empirical equation for the variation of C_p, to add to Debye's or a similar equation an extra term $aT^{3/2}$, which is always only small, so that we obtain

$$C_p = f\!\left(\frac{\beta \nu}{T}\right) + aT^{3/2}; \quad . \quad . \quad . \quad (21)$$

the first term on the right-hand side of this equation gives at high temperatures the limiting value $3R = 5\cdot955$, while in practice an increase above this theoretical limit is always observed; the increase, especially when not too close to the melting-point, amounts only to a few tenths, as is also to be seen from the observed curves of Fig. 10. In this figure a suitable extra term has been included in the theoretical curves.

6. Determination of Atomic Weights from Specific Heats.—Both the law of Avogadro and that of Dulong and Petit have, of course, been much used in the determination of atomic weights; originally their use was only for an approximate estimation, while the exact value was then found by chemical analysis.

In the last decades Ph. A. Guye, in Geneva, and D. Berthelot, in Paris, have been able to make extremely accurate and important atomic weight determinations by reducing the observed densities of gases to the ideal gaseous state; these determinations could compete with the best analytical measurements.

It is sure to be of general interest to point out that now that we have given an exact formulation of the law of Dulong and Petit, it seems that a very accurate determination of atomic weight may sometimes be possible by this method.

As an example, let us deal with silver, which is conveniently obtainable in an extremely pure state, and for which particularly reliable measurements are consequently available. In order to derive the atomic weight, the measured specific heat must be reduced to constant volume, and then that factor must be found which, multiplied by the specific heat, best reproduces the variation required by equation **(18)**; this factor is then equal to the atomic weight.

The calculation made by F. A. Lindemann gives for silver the values :—

LAW OF DULONG AND PETIT

	M =	106·8	107·3	107·88	108·42	108·96
Most probable value of $\beta\nu$		223	222	221	221	217
Sum of the squares of the errors		5·07	4·88	5·05	6·78	9·05

The most probable value for M is, of course, that corresponding with the minimum sum of the squares of the errors: the atomic weight obtained by plotting the above values is

$$M = 107\cdot55 \text{ instead of } 107\cdot88.$$

From the practical point of view, we must not give undue weight to this good agreement, which is probably to be ascribed to the fact that exact results have been obtained for the specific heat of silver by different observers; but it may well serve to indicate afresh that the theoretical foundations for the law of Dulong and Petit are now just as well established as those for the law of Avogadro have been for many years.

One might in practice proceed in the following simple manner. Suppose the specific heat is measured with the highest accuracy at one temperature in a region where the specific heat is only slowly increasing, i.e. where the law of Dulong and Petit is nearly obeyed. Let this be reduced to constant volume (which constitutes the chief difficulty). Then let the theoretical atomic heat, which will lie only a little below 3R, be calculated for that temperature by using a value of ν which is only approximate, having been obtained from a single measurement at low temperature, or from Lindemann's equation; division of these quantities one into the other gives the atomic weight. Similarly, of course, compounds could be used instead of the solid element.

7. Estimation of the Values of ν.—The experimental and theoretical investigations described in this and the previous chapter have apparently led to a complete explanation of the law of Dulong and Petit, and of its generalization by Kopp and Neumann. A knowledge only of the one value of $\beta\nu$, for very simply constructed crystals, or of the various values of $\beta\nu$ for crystals of complicated structure, is required to give us a conspectus of the whole variation of the atomic and molecular heats, and to indicate, therefore, the range of temperature in which the above laws hold.

Now there are various methods of estimating the value of $\beta\nu$ with more or less accuracy, at least for simply constructed substances, among which are included practically all metals. Any of these methods allows us to predict to what extent the substance in question obeys the above laws at a given temperature; this is further evidence that one may consider these laws to have been fully explained.

This is not the place for consideration of these methods, so that we shall content ourselves with enumerating the most important; this enumeration may show us the diversity both of the measurements, which allow the behaviour of a solid towards the Dulong and Petit law to be foreseen, and of the types of phenomenon which may be brought by its means into one common field of view.

Values of $\beta\nu$ have been successfully calculated from Rubens' "reststrahlen" (Einstein, Nernst, and Lindemann), from the elastic properties (Sutherland, Einstein, Debye, Born and v. Karman, Eucken), from the electrical conductivity of metals (Nernst, Kamerlingh-Onnes, Schiemank), and from thermal expansion (Grüneisen).*

The development of methods of this sort for the indirect determination of values of $\beta\nu$, and hence of the energy content of solids (even amorphous), is naturally of special importance for the purpose here in hand, for all applications of the Heat Theorem depend, in the last resort, on a knowledge of the energy content. Special attention may be drawn to an original research made with this object (Koref, 76).

* I have given a review in four lectures on "The Theory of the Solid State," delivered at the University of London (University of London Press, 1914).

CHAPTER V

SPECIFIC HEATS OF GASES

1. General.—It is a question of fundamental importance for the applications of thermodynamics, and of my Heat Theorem in particular, whether we know the value of the integral

$$E = \int_0^T C_p \, dT \qquad \qquad \textbf{(22)}$$

for gases; we have seen in the foregoing chapters that this question has received an experimental solution for all crystalline bodies and for liquids (amorphous bodies), at least in certain cases, and that there is weighty theoretical assistance from the T^3-law of Debye, *inter alia*, in support of experiment.

Even though we have yet no final solution in the realm of gaseous bodies, we have in the last few years made great advances, some of which have led to totally new ideas. In particular, the phenomenon of " degeneration (Entartung) of gases " is among the greatest surprises which recent advances in natural science have produced.

Before going into details I should like to refer briefly to two results which are of importance for our purposes :—

1. All gases, at sufficiently low temperatures and under sufficiently low pressures, behave like monatomic gases as regards their specific heats, i.e. their molecular heat at constant volume converges towards the value $\frac{3}{2}R$.

2. If we suppose a gas of finite concentration to be cooled at constant volume, avoiding any condensation, it passes into a state totally different from the ordinary gaseous state

in which its pressure becomes independent of temperature, and its specific heat negligibly small (the degenerated state); but for rarefied gases this region lies very close to the absolute zero.

2. Reduction to the Ideal Gaseous State.—In order to avoid complications, which, though appearing in the liquid state, are much more important with gases, we must reduce the specific heat of gases, whether measured at constant pressure or at constant volume, to the ideal gaseous state. This can be done by means of the two equations, derived from the Second Law :—

$$C_p - C_v = T\frac{\partial p}{\partial T} \cdot \frac{\partial v}{\partial T} \quad . \quad . \quad . \quad (23)$$

$$\frac{\partial C_v}{\partial v} = T\frac{\partial^2 p}{\partial T^2} \quad . \quad . \quad . \quad (24)$$

It is most simple to use the excellent equation of state due to Daniel Berthelot.* For ideal gases, for which

$$pv = RT$$

holds, we have, of course,

$$C_p - C_v = R = 1 \cdot 985 \text{ cals.} \quad . \quad . \quad (25)$$

We shall not go into details in this connection, and I shall only emphasize that, for exact theoretical investigations, reduction to the ideal gaseous state is always necessary and can, moreover, always be done quite safely. Molecular heats· at constant volume and constant pressure may be calculated, by means of Berthelot's equation, for any density which is not too great, from the critical pressure and the critical temperature if there is a reliable measurement available at a particular pressure for the required temperature.

All subsequent considerations of specific heat in this book concern the values calculated to the state of ideal gas; for rarefied gases this recalculation is superfluous.

* Cf. " Theor. Chem.," p. 256.

SPECIFIC HEATS OF GASES

3. Experimental Methods.—Here again I must be quite brief and must refrain from dealing fully with the work of my pupils Voller (18), Levy (32), Keutel (44), Thibaut (54) and Partington (88), or with my own work on the specific heats of gases. I shall refer to two methods only, of which one is useful for very high, the other for very low temperatures. Pier (43) succeeded, after several years of effort, in rendering much more refined the explosion method used by Bunsen, le Chatelier, and others; Bjerrum, Siegel, and especially K. Wohl, who continued the work, made further improvements and, in particular, extended it to about 3000°. We have thus quite an accurate idea of the variation of specific heat for a number of gases up to very high temperatures. For very low temperatures Eucken (105) has worked out a method which gives the specific heat of a gas at constant volume; he uses the vacuum calorimeter (p. 48), and relies on the fact that at very low temperatures the heat capacity of a steel or copper vessel containing the gas to be examined becomes extremely small. This method breaks down, of course, for temperatures at which the vapour pressure of the gas in question falls to an atmosphere or less; to supplement it a method which allowed the specific heat of a gas to be measured at very low pressure (e.g. 0·001 mm.) would be very valuable, and seems to be quite feasible.

In the table * on next page are given some results of our measurements of molecular heat at constant volume, using, of course, also the results of the older researches of Regnault, E. Wiedemann, Strecker, Holborn, and Henning, etc.

For low temperatures Eucken found for hydrogen, reduced to the state of an ideal gas—

$T = 35°$ $50°$ $80°$ $100°$ $273°$
$C_v = 2·98$ $3·01$ $3·14$ $3·42$ $4·84$

For helium at a concentration of 30 mols per litre—

$T = 18°$ $22°$ $26°$ $30°$
$C_v = 2·90$ $3·00$ $3·10$ $3·10$

* See also Supplement, p. 265.

TABLE V
Molecular Heats of Gases

Gas.	Value for a Rigid Molecule.	Temperature °C.					
		0	100	200	500	1200	2000
A, I	2·978	2·98	2·98	2·98	2·98	3·0	3·0
H_2	4·963	4·87	4·93	5·04	5·16	5·67	6·28
N_2, O_2, CO	4·963	4·99	5·05	5·15	5·26	5·75	6·3
HCl	4·963	5·00	5·00	5·27	5·46	6·13	6·9
Cl_2	4·963	5·95	6·3	6·7	6·9	7·1	7·2
H_2O	5·955	5·93	6·00	6·60	7·00	8·4	11·0
CO_2	5·955	6·68	7·69	9·04	9·75	10·6	11·1
SO_2	5·955	7·2	8·1	(9·2)	(9·8)	(10·6)	(11·1)
NH_3	5·955	6·62	7·05	8·3	9·5	11·4	—
$(C_2H_5)_2O_5$	5·955	ca. 30	33	42	—	—	—

4. Application of the Quantum Theory.—The classical kinetic theory of heat has thrown light on many important points in regard to specific heat, not only, as we saw on page 57, for solid bodies, but also, and in a greater degree, for gases. As with solids, deviations were encountered in the case of gases which defied elucidation.

The postulates of the older theory may be summarized as follows: if n be the number of degrees of freedom in the molecule, then we have simply

$$C_v = n\frac{R}{2}.$$

According to this a monatomic gas, with its three degrees of freedom for linear motion, must have the molecular heat $\frac{3}{2}R$; this is, of course, true, and is always rightly reckoned as one of the finest results of the kinetic theory. Diatomic rigid molecules have five, triatomic rigid molecules six degrees of freedom, since in the first case there are two, in the last case three, more degrees of freedom of rotation. As the above table shows, there occurs in many cases a remarkable approximation to the values

$$C_v = \tfrac{5}{2}R \text{ or } \tfrac{6}{2}R.$$

SPECIFIC HEATS OF GASES

Finally, there are possible also vibrations of the atoms within the molecule, which are apparently to be treated just like the vibrations of atoms in the crystal. These explain the occurrence of molecular heats higher than $\frac{5}{2}R$ for diatomic, or than $\frac{6}{2}R$ for polyatomic gases.

On closer examination, however, several doubtful points arise: in particular, the older theory has difficulty in explaining the gradual rise of specific heat at higher temperatures, which is always observed in practice. I was able to show (51) in 1911 that application of the quantum theory not only overcomes these difficulties, but may lead to totally new points of view. I pointed out also at the Solvay Congress (1911) that even the conceptions with which the kinetic theory supplies us for a monatomic gas cannot be completely satisfactory, and that quite a different state of affairs must exist, particularly at very low temperatures (cf. also Nernst, 47 and 66).

With the aid of the quantum theory we may make the following general picture of the behaviour of the specific heats of gases.

Consider a molecule of gas consisting of any number of atoms: at a sufficiently high temperature, energy will be stored up not only in the translation and in the rotation of the molecule, but also internally, in the vibrations of the atoms to and fro. The latter amount is obviously to be calculated just as was done on page 60 for solids, and it must disappear at sufficiently low temperatures for the same reasons as the specific heat of solids there converges towards zero.

But according to the quantum theory even the rotational movements of the molecule cannot take place as was supposed by the older kinetic theory. They, too, must become negligible at sufficiently low temperatures, i.e. must contribute no appreciable amount to the specific heat, for otherwise the thermal motion would be in disagreement with the radiation.

Having drawn this conclusion, I showed by rough calculation that this quantum effect should appear in the case of hydrogen at temperatures which could be attained, and that it was evinced, though only feebly, even in Regnault's measurements carried out at room temperatures. The measurements by Eucken, above mentioned,* then proved that at low temperatures hydrogen assumes the same molecular heat as monatomic gases.

In these circumstances it has very reasonably been assumed by all theorists who have occupied themselves with the question, that this behaviour is general or, in other words, that at sufficiently low temperatures C_v for all gases converges towards the limit $\frac{3}{2}R$, i.e. towards the same value which is given by monatomic gases even at higher temperatures.

We shall see later (Chapter XIII) that this result is of the greatest importance in thermodynamics. We shall now deal briefly with the "degeneration of monatomic gases"; the following considerations apply, of course, also to gases which have been so cooled as to lose both their energy of internal vibration and their energy of rotation, and have thus been brought into the thermally monatomic state, as we may call it for the sake of brevity.

5. Monatomic Gases.†—All measurements, from the time of Kundt and Warburg until now, agree here in finding always that

$$E_v = \tfrac{3}{2}RT \qquad . \quad . \quad . \quad (26)$$

and hence

$$C_v = \tfrac{3}{2}R = 2\cdot978 \qquad . \quad . \quad . \quad (27)$$

Table V (p. 70) confirms this result for argon and iodine up to very high temperatures. This result is, of course,

---

* Scheel and Heuse ("Ann. d. Physik," **40**, 473, 1913) have obtained a practically identical result by a totally different method, though they did not go down to such low temperatures as did Eucken.

† Nernst (66).

SPECIFIC HEATS OF GASES

obtained if we apply the law of equipartition of energy to a non-rotating point mass; we have then three degrees of freedom, each of which has an amount of energy $\frac{R}{2}T$. We cannot, however, at the present time regard equation **(26)** as quite exact; for the impact of two atoms, which we may even suppose electrically charged, must of necessity be associated with alterations in the direction of their previous translation, and must, therefore, cause a radiation to be emitted. Of necessity, therefore, at low temperatures equations analogous to **(12)** or **(18)** must hold instead of equation **(26)**. The fact that equation **(27)** holds with almost perfect accuracy at ordinary temperatures is, of course, compatible with this. A method for theoretical calculation will be described in Chapter XIV. It has been suggested to me verbally by friends that, to a first approximation, the number of impacts of an atom with other atoms may be assumed as the value of ν in equations **(12)** or **(18)**. This assumption does not agree with the facts; for in the case of hydrogen at atmospheric pressure and ordinary temperatures, this would give

$$\nu = \frac{183900}{0\cdot 0000185} = 10^{10}; \quad \beta\nu = 0\cdot 48°.$$

This result would mean that on expanding hydrogen, compressed, say, to 100 atmospheres where $\beta\nu$ would be equal to 48°, a cooling of more than 10° must take place. On the contrary there is, in these circumstances, at ordinary temperature a development of heat, though an extremely small one. We can also show by another method that the law of equipartition of energy

$$\frac{m_1 u_1^2}{2} = \frac{m_2 u_2^2}{2}$$

(m = mass, u = mean velocity of the atom) must hold very accurately for the energy of translation in gases, since this equation conditions the validity of Avogadro's law. Now the law has been very accurately verified, with the aid of chemical analysis, by the work of D. Berthelot, in Paris, and of Ph. Guye, in Geneva. We may therefore conclude that if we imagine hydrogen or oxygen, say, under normal conditions to be cooled (in the superheated state, of course, at low temperatures) to the absolute zero, no decrease of specific heat due to loss of energy of translation can result until the temperature is so low that equation **(26)** may be regarded as true to within about one part in a thousand, and hence equation **(27)** only becomes invalid a few tenths of a degree above the absolute zero. Accordingly, we may regard the question of energy content as settled with very great accuracy for monatomic gases. Nevertheless, a theory of the decrease of specific heat which, as we have just mentioned, is to be expected at very low temperatures, is of very great practical importance, for such a theory leads to a vapour-pressure formula, and hence also to a theoretical determination of the so-called chemical constants of gases. In Chapter XIV we shall deal more fully with the researches made up to date with this object.

From the experimental side we have already a certain confirmation of the theory of " degeneration " in the circumstance (cf. p. 71) that C_v for compressed helium assumes a value less than 2·98 ; at any rate, such a behaviour was not to be expected for gases on any of our previous conceptions.

6. Summary.—Let us, in conclusion, picture to ourselves how a polyatomic gas like ammonia behaves on cooling.

At high temperatures we must expect a very high molecular heat on account of the energy of the internal vibrations ; with moderate cooling, since fairly firm compounds and consequently relatively high values of ν are in question, this energy falls to a very small value. Even at room temperature, where the molecular heat of ammonia, according to Table V (p. 72), amounts to 6·7, the value for a rigid molecule ($\frac{6}{2}R = 5·96$) is almost reached, and it should be fairly accurately attained in the neighbourhood of T = 200°. On still further cooling, the energy of rotation also begins to disappear; measurements are not available here, and would be difficult to make on account of the extremely small vapour pressure of ammonia in this region of temperature, but it can no longer be doubted that the molecular heat converges to $C_v = \frac{3}{2}R$.

Provided we keep near to the state of an ideal gas, the specific heats are independent of the concentration; according to the theory this ceases to be the case when we reach the region of very low temperatures, i.e. the region of "degeneration." The higher the concentration the earlier does this last decrease leading to the value $C_v = 0$ begin.*

The points of view set out in this chapter, which obviously take us far beyond the older kinetic theory of the specific heats of gases, have brought new light in various directions; for these we are indebted solely to the quantum theory, which has here shown most clearly its heuristic value.

---

* The circumstance, that in the region of degeneration (and only there) the specific heat, even of very rarefied gases, depends on the density, is of great importance thermodynamically, as we shall explain in Chapter XIV.

CHAPTER VI

FORMULATION OF THE NEW HEAT THEOREM

Formulation of the new Heat Theorem.—We begin with the equation on the basis of which may be derived all the applications of the Second Law hitherto made (cf. p. 3),

$$A - U = T\frac{dA}{dT} \quad . \quad . \quad . \quad . \quad (1)$$

For example, Helmholtz's important application to galvanic cells runs—

$$EF - U = TF\frac{dE}{dT} \quad . \quad . \quad . \quad (28)$$

(F = electro-chemical equivalent.) Equation **(1)** has been used throughout several decades without any objections being raised against it; strangely enough, scruples about this equation have only arisen since I extended it. These are, of course, due to misunderstandings, and we are not here concerned to clear them up, since we must assume a knowledge of classical thermodynamics.

In order to eliminate such misunderstandings from the outset, let us assume that there is a natural process to which the application of the equation

$$A - U = T\frac{dA}{dT} \quad . \quad . \quad . \quad . \quad (1)$$

has been made without objections; this assumes that the question of the nature of the two thermodynamical functions A and U has been made quite clear, which, of course, presupposes certain knowledge of the type of the natural process considered (thus, for example, the application to galvanic cells would not be possible without a knowledge of Faraday's law).

FORMULATION OF NEW HEAT THEOREM

The First Law shows us, according to Kirchhoff,*

$$\frac{dU}{dT} = C - C' \quad . \quad . \quad . \quad (29)$$

(C is the heat capacity of the system before, and C' that after the change.)

We shall now assume that the system under examination can be cooled to the absolute zero continuously, i.e. without any part of the system undergoing a complete change at one temperature, such as freezing of a liquid, transformation of one modification into another, etc. This assumption only requires, thermodynamically, that *with any small lowering of temperature there is always associated only a very small abstraction of heat*, and that *this is true right down to the absolute zero*. Now the fact that the specific heats of all parts of the system are continuous functions of temperature, means that they may therefore be expressed, with an approximation sufficient for any attainable accuracy of measurement, by a series ascending in whole powers of T.

Equation (29) then shows at once that U may also be brought into the form

$$U = U_0 + \alpha T + \beta T^2 + \gamma T^3 + \ldots \quad . \quad (30)$$

If we introduce this into the equation

$$A - U = T\frac{dA}{dT} \quad . \quad . \quad . \quad (1)$$

and integrate, we obtain, as may be readily verified by substitution in (1),

$$A = U_0 + a_0 T - \alpha T \log_e T - \beta T^2 - \frac{\gamma}{2}T^3 \quad . \quad (31)$$

where a_0 is an unknown constant of integration.

By differentiation,

$$\frac{dU}{dT} = \alpha + 2\beta T + 3\gamma T^2 + \ldots \quad . \quad . \quad (32)$$

$$\frac{dA}{dT} = a_0 - \alpha \log_e T - \alpha - 2\beta T - \tfrac{3}{2}\gamma T^2 - \ldots \quad . \quad (33)$$

* Cf. e.g. " Theor. Chem.," p. 688.

It follows from the last equation that even though A itself (equation **31**) passes continuously into the value U_0 at the absolute zero, $\frac{dA}{dT}$ then becomes infinitely great.

This result does not agree with expectations, for it might have been supposed that A, like U, would show a simple behaviour at very low temperatures. It is pertinent to ask what the circumstances must be which are required to obviate this disturbing behaviour.

The answer is simply that α must be equal to nil (not merely that α must have a " small " value, which assumption would be of no use to us). This requirement may also be written

$$\lim \frac{dU}{dT} = 0 \quad \text{for } T = 0 \quad . \quad . \quad (34)$$

If we make the further assumption that at very low temperatures A and U behave similarly, i.e. that we must also have

$$\lim \frac{dA}{dT} = 0 \quad \text{for } T = 0 \quad . \quad . \quad (35)$$

(which is identical with the assumption $a_0 = 0$), we see that the validity of equation **(34)** is a consequence of **(35)** (but not *vice versa*). For if we put the right side of equation **(33)** equal to 0 for very low temperatures, it follows at once that $a = 0$ and $a_0 = 0$.

Equation **(35)** is the complete content of the new Heat Theorem, and all the applications hitherto made follow from it.

The following reflections show how far-reaching this equation is. It is not possible to calculate A from the equation

$$A - U = T\frac{dA}{dT} \quad . \quad . \quad . \quad (1)$$

even when U is known as a function of the temperature, since on integrating this differential equation an unknown constant of integration appears; this, though independent of temperature, varies from case to case, and must be determined afresh for each new special application.

But if we combine with it

$$\lim \frac{dA}{dT} = 0 \quad \text{for } T = 0 \qquad . \qquad . \quad (35)$$

it is possible to calculate A; if U and A have the special form of equations (30) and (31) we obtain, since as above $a = 0$ and $a_0 = 0$, the simple results,

$$U = U_0 + \beta T^2 + \gamma T^3 + \ldots \qquad . \quad (36)$$

$$A = U_0 - \beta T^2 - \frac{\gamma}{2} T^3 - \ldots \qquad . \quad (37)$$

i.e. knowing U_0, β, γ, ..., we can calculate not only U but also A.

If we integrate equation (1) generally we find

$$A = -T \int \frac{U}{T^2} dT + a_0 T \qquad . \qquad . \quad (38)$$

and by combination with (35)

$$A = -T \int^T \frac{U}{T^2} dT \qquad . \qquad . \quad (39)$$

This method of writing indicates that the expression following the integration sign is to be integrated indefinitely, and the upper limit to be inserted.

We see at once that A may be calculated from purely thermal data, which was not possible when using the hitherto known laws.

A slightly different formulation of the new Heat Theorem states

$$\lim \frac{dA}{dT} = \lim \frac{dU}{dT} \text{ for } T = 0 \qquad . \qquad . \quad (40)$$

As follows from the two equations (32) and (33), this rule also leads to the results

$$a = 0 \quad \text{and} \quad a_0 = 0,$$

i.e. to the formulæ **(36)** and **(37)**. But the formulation **(40)** is particularly illuminating, in that it requires that at low temperatures A and U shall become tangential, i.e. follow an identical course ; this explains the fact, mentioned in the introduction, that at low temperatures the Principle of Berthelot holds remarkably well for condensed systems.

It was the conception that Berthelot's Principle (with its consequences) was a limiting law, valid for condensed systems not only at the absolute zero, but also in its neighbourhood, which led me to propound equation **(40)**, and thus to discover the new Heat Theorem. But this conception demanded the previous recognition of the fact that it was useless to apply Berthelot's Principle to gases immediately (i.e. without the idea of " degeneration ").

The relation

$$\lim \frac{dU}{dT} = 0 \quad \text{for } T = 0.$$

shows also that at low temperatures the atomic heats of elements and compounds must be strictly additive. In the first practical applications of my Theorem I was led to the conception that they must all converge at low temperatures to very small values. The experimental and theoretical work of later years has not only confirmed this idea, but has made it practically certain (Chapter IV) that *at low temperatures the specific heats of all bodies converge towards zero.*

We shall make use of this in the sequel, and by taking into account the very important result (p. 61), obtained theoretically by Debye and confirmed experimentally by Eucken and more recently by Schwers and myself, according to which the specific heat at low temperatures varies as the third power of the absolute temperature, we shall then find

$$U = U_0 + \delta T^4 \quad \text{and} \quad A = U_0 - \frac{\delta}{3} T^4 \quad . \quad (41)$$

as equations *strictly* valid for low temperatures.

Graphical representation has played a great part in the

FORMULATION OF NEW HEAT THEOREM

development of thermodynamics; I need only recall the cyclic processes used by Carnot, Clapeyron, and Clausius. We shall therefore briefly recapitulate the above observations by making use of some simple diagrams.

The general integral of the equation

$$A - U = T \frac{dA}{dT} \qquad \qquad (1)$$

is
$$A = - T \int \frac{U}{T^2} dT + a_0 T; \qquad (38)$$

when $T = 0$,
$$A = U_0,$$

i.e. as above explained, the law of Berthelot is valid at the absolute zero without limitations.

The new Heat Theorem, according to which a_0 must be equal to 0, teaches us that the curves for A and U not only meet at the absolute zero, but touch one another before they reach it: according to Debye's law this contact is very close, being of the third degree.

As above shown, the common tangent to the two curves must be parallel to the axis of abscissæ.

To fix our ideas we shall now assume that we are given the value of $U = U'$ for a given temperature T', as a result, perhaps, of a calorimetric determination. Also, let all the specific heats required be known down to very low temperatures. Then, by Kirchhoff's law, we can also deduce U down to very low temperatures, since the direction in which U must be drawn is given, according to the equation

$$\frac{dU}{dT} = C - C' . \qquad (29)$$

If the measurements of specific heat have been extended to temperatures so low that C and C' have fallen to very low values, and thus $C - C'$ is also only very small, we can simply draw the last part of the curve as parallel to the axis of temperature without occasioning any appreciable error; but even if this is not the case we can usually, e.g. by use

84 THE NEW HEAT THEOREM

of the T^3-law, undertake the extrapolation of U to the absolute zero without any considerable uncertainty.

As already set out on page 81, it is now possible to draw the A-curve in full and with complete certainty. The first part of this curve, starting from the absolute zero, we again trace conveniently with the aid of the T^3-law (by equation **41**), while at higher temperatures we continue A in the direction given by the equation

$$\frac{dA}{dT} = \frac{A-U}{T}.$$

The following geometrical artifice discovered by Gans * and by Drägert (102) considerably facilitates this operation.

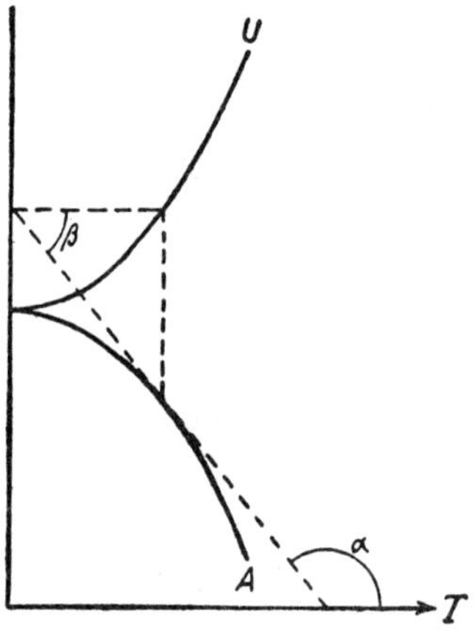

From a point on the U-curve (Fig. 11) draw lines parallel to the ordinate and to the abscissa, and from the point of intersection of the former with the A-curve draw a straight line to the intersection of the latter with the ordinate. The triangle so produced gives us

$$\tan \beta = -\tan \alpha = S_2 - S_1 = \frac{Q}{T}$$

FIG. 11.

i.e. the straight line above found is the tangent to the A-curve.

By obtaining successive tangents to the A-curve the latter may be drawn in full, starting from U_0; this can of course only be done if the original direction of the A-curve, starting from the point U_0, is regarded as fixed with the aid of the Heat Theorem (cf. Fig. 1, p. 12).

For the sake of completeness, the formulation which our Heat Theorem takes when dealing with entropy may be given.

* Cf. further, Chapter XVI.

The entropy S is defined thermodynamically by the equation

$$dS = \frac{dQ}{T}$$

where dS is the increase of entropy in a process occurring at the temperature T with the absorption of heat dQ; on integration we obtain

$$S_2 - S_1 = \int_1^2 \frac{dQ}{T}.$$

For an isothermal reversible process it follows that

$$S_2 - S_1 = \frac{Q}{T} = \frac{A - U}{T},$$

and, making use of the equation **(1)**,

$$S_2 - S_1 = \frac{dA}{dT}.$$

Our Heat Theorem therefore takes the form

$$\lim (S_2 - S_1) = 0 \quad \text{for } T = 0;$$

thus it states that *in the neighbourhood of the absolute zero all processes proceed without alteration of entropy.* Applied to the case of transformation of allotropic modifications of a substance into one another, this means that the entropy of both forms is the *same* at the absolute zero.

We may now go a step further and, with Planck,* put the separate entropies equal to zero. Then our Heat Theorem states: *At the absolute zero of temperature the entropy of every chemically homogeneous solid or liquid body has a zero value* The entropy of a body at the temperature T is thus

$$S = \int_0^T \frac{dQ}{T} = \int_0^T \frac{c\,dT}{T}.$$

---

* Planck, "Thermodynamik," Leipzig, 1911, p. 269

Then, since at finite temperatures the entropy has finite values, the molecular heat at the absolute zero must also be negligibly small; hitherto, with the equation

$$\lim \frac{d\mathrm{U}}{d\mathrm{T}} = 0 \quad \text{for } \mathrm{T} = 0,$$

we have only required that near the absolute zero the heat capacity should remain unchanged in any transformation.

The assumption that all specific heats (even those of gases) become negligibly small near the absolute zero appears to have been so well established both by theory and also, in so far as it was possible, by experiment, that no serious objection is now likely to be raised against it. Thus every distinction between the above conceptions has now disappeared; for the practical applications of the Heat Theorem, in which only changes of entropy are concerned, none has ever existed.

CHAPTER VII

PRINCIPLE OF THE UNATTAINABILITY OF THE ABSOLUTE ZERO.

1. General.—From the consideration of a cyclic process we may prove the following law :—

There cannot be any process taking place in finite dimensions by means of which a body can be cooled to the absolute zero.

This law we shall call the *Principle of the Unattainability of the Absolute Zero* (65).

The proof is as follows :—

If the above law did not hold we could suppose that there was a cyclic process in which the lower isotherm DC (cf. the diagram below, which is drawn in the usual manner) is at the absolute zero. We might, for instance, suppose that a solid or liquid body expanded at the low but finite absolute temperature ΔT, while it was maintained in constant communication with a heat reservoir—isotherm AB; that it was then disconnected from the reservoir and further expanded adiabatically until it became cooled to the absolute zero—adiabatic BC. It is then compressed at the absolute zero—isothermal CD—and finally warmed an infinitely small amount by an infinitely small addition of external work, such that it is brought again by the adiabatic compression to the temperature ΔT and the original volume—adiabatic DA.

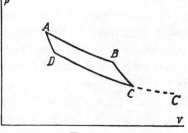

Fig. 12.

Now in such a reversible cyclic process a certain finite amount of external work will be supplied, which is given by the area contained between the four curves. Since no heat is taken up on the curves BC and DA, and since, by equation (1), no heat can be taken up over the curve CD because the latent heat disappears at the absolute zero, it follows that the external work must have been supplied at the cost of the heat removed from the heat reservoir at the temperature ΔT, which is very low perhaps, but still finite. But as this contradicts the Second Law, we arrive at the conclusion which was to be demonstrated.

The above proof is perfectly rigid; the only assumption it contains is that at the absolute zero any reversible change of state is necessarily isothermal, i.e. that the relation holds,

$$A - U = T\frac{dA}{dT} = 0 \quad \text{for } T = 0 \qquad . \quad . \quad (1)$$

This is, however, a supposition which has always been made from the standpoint of classical thermodynamics, and it has never hitherto been, nor can it well be, controverted

The objection that it is not possible to go from point C along the curve CD, but that the curve CB must be retraced, is based on a fundamental misunderstanding. For if $T = 0$ has been reached, and that, *ex hypothesi*, not asymptotically but directly by a " process occurring in finite dimensions," one must of necessity, on reversing the process (in the above example on compressing the body), pass from C to D and not to B, because of the identity of A and U at the absolute zero.

In order to return from D to A adiabatically, on the contrary, we must, as mentioned above (and also by me in the paper cited on p. 87), warm the body an infinitely small amount, so that it leaves the absolute zero, and becomes capable of adiabatic compression.

The following reasoning also shows that the portion of

the curve BC is itself contrary to nature. If we allow the body to expand further after it has reached the state C it would follow the curve CC'; in other words, it would show the paradoxical behaviour of expanding first from B to C adiabatically, and then from C to C' isothermally.

This absurdity disappears if we regard the curve BC as impossible, i.e. if the principle first enunciated is true; it remains to be found under what conditions the curve BC would be realizable, and these conditions are then to be regarded as contradicting a general law of nature.

The reason why the principle has never been particularly emphasized hitherto is probably that, on the usual older assumption that the specific heat remains finite, however low the temperature, it seemed to be obvious that it would not be possible to reach the absolute zero. Since, as already mentioned, latent heats become very small at very low temperatures according to the Second Law, no further calculation appeared to be required to show that it was impossible to abstract from a body its heat content, which was finite even at the lowest finite temperature, and approximately proportional to the temperature.

The question assumes a totally different aspect when we make the supposition that, in the general expression for specific heat,

$$C_v = c_0 + a\mathrm{T} + b\mathrm{T}^2 + c\mathrm{T}^3 + \ldots \qquad \textbf{(42)}$$

not only c_0, but also the coefficients of the two following terms become negligibly small. As a matter of fact, we shall manage, in what follows, with the assumption that only the first term on the right-hand side of equation **(42)** disappears; if higher coefficients, which are multiplied by finite powers of T, are negligibly small, the following results hold *a fortiori*.

We shall carry out the calculation of the above described cyclic process for two special examples, viz. the expansion of a solid and a chemical process.

2. Expansion of a Solid.

—Our fundamental formula

$$A - U = T\frac{dA}{dT} \quad . \quad . \quad . \quad (1)$$

becomes, if $-U$ denote the internal energy of the body considered, p and v the pressure and the volume,

$$p - \frac{\partial U}{\partial v} = T\frac{\partial p}{\partial T} \quad . \quad . \quad . \quad (43)$$

The heat introduced amounts, according to the First Law, to

$$Q = C_v dT - \frac{\partial U}{\partial v}dv + pdv.$$

Combining the last two equations, we have for adiabatic compression or dilatation ($Q = 0$),

$$0 = C_v dT + T\frac{\partial p}{\partial T}dv.$$

If we generalize the coefficient of pressure as

$$\frac{\partial p}{\partial T} = a_0 + a_1 T + a_2 T^2 + \ldots \quad . \quad . \quad (44)$$

it follows that

$$-\frac{dT}{T} = \frac{a_0 + a_1 T + \ldots}{aT + bT^2}dv.$$

Thus for low temperatures

$$-dT = \frac{a_0}{a}dv,$$

i.e. even at the lowest temperatures a finite change in volume entails a finite cooling; the change in volume is, of course, supposed to occur in the sense in which it is associated with absorption of heat. But this can only signify that the attainment of the absolute zero should be possible.

This result is prevented if

$$a_0 = 0 \quad . \quad . \quad . \quad . \quad (45)$$

UNATTAINABILITY OF ABSOLUTE ZERO

for then
$$\Delta v = \frac{a}{a_1} \log_e \frac{\Delta T}{T} \quad . \quad . \quad . \quad (46)$$

i.e. an infinitely great change in volume is required to pass from a low but finite temperature ΔT to the absolute zero, $T = 0$.

Now,
$$\lim \frac{\partial p}{\partial T} = a_0 \text{ for } T = 0 \quad . \quad . \quad (47)$$

thus the requirement (45) is identical with that of the Heat Theorem (cf. Chapter IX).

3. Chemical Change.—If it be assumed that for each of the reacting substances
$$c_0 = 0,$$
equation (1) gives for the latent heat, at sufficiently low temperatures,
$$A - U_0 = a_0 T$$

(a_0 being the constant of integration). To simplify the method of expression, let us consider any simple case, say the transformation of graphite into diamond, and assume, in order to fix our ideas, that a_0 is a positive quantity. Then there is associated with the reversible change dv the heat absorption $a_0 T dv$, and we get for the adiabatic transformation
$$a_0 T dv + (aT + bT^2 + \ldots)dT = 0,$$
or, integrating, at sufficiently low temperatures,
$$\Delta v = \frac{a}{a_0}(T_2 - T_1).$$

Thus the absolute zero would be reached with an infinitely small transformation dv if the original temperature were dT.

If, according to our Heat Theorem, we put
$$a_0 = 0,$$

we get the equations
$$A = A_0 - aT^2,$$
$$U = A_0 + aT^2;$$

thus the latent heat at low temperatures would be
$$A - U = -2aT^2,$$

and consequently, as above,

$$\Delta \nu = -\frac{a}{2a} \log_e \frac{T_2^2}{T_1} \quad . \quad . \quad . \quad (48)$$

i.e. the absolute zero ($T_2 = 0$) is not to be reached by the change of a finite amount of the substance.

In a similar manner a proof may be given, for any process whatsoever, that, on the assumption that the specific heats become negligibly small, the requirement that the curve BC shall be incapable of being realized coincides with the conclusions to be drawn from our Heat Theorem.

Since, on the other hand—unless the absolute zero be arbitrarily excluded from thermodynamical treatment—classical thermodynamics requires the cyclic process first described to be unrealizable, we must conclude that the Heat Theorem is theoretically demonstrated for a system whose heat capacity vanishes in the neighbourhood of the absolute zero.*

The enunciation of the First and Second Laws may be ascribed to the observation that certain arrangements cannot be realized in spite of all endeavours. Similarly, the new Heat Theorem, had it not been found in this somewhat roundabout way, may be recognized in what is apparently its most general outline by the impossibility of obtaining a certain effect. We can thus embrace the three Laws of Thermodynamics now known in the following theses :—

(1) It is impossible to construct a machine which con-

---

* The above proof has been given in a slightly different form by Polanyi: we shall return to it in the last chapter.

UNATTAINABILITY OF ABSOLUTE ZERO

tinuously produces heat or external work out of nothing.

(2) It is impossible to construct a machine which continuously converts the heat of its surroundings into external work.

(3) It is impossible to devise an arrangement by which a body may be completely deprived of its heat, i.e. cooled to the absolute zero.

From the first proposition may be deduced every consequence which the Law of the Conservation of Energy entails. From the second proposition we can deduce the correctness of the fundamental equation of the Second Law of Thermodynamics, if we introduce the concept of temperature. From the third proposition we can derive the mathematical formulation of the new Heat Theorem, if we make use of the observation that specific heats become negligibly small at very low temperatures.

CHAPTER VIII

SOME IMPORTANT MATHEMATICAL FORMULÆ

IN the application of the Heat Theorem we are mainly concerned with the evaluation of the following integrals:—

$$E = \int_0^T C_p dT \text{ and } F = T\int_0^T \frac{E}{T^2} dT.$$

If, as is always possible and generally necessary, we develop C_p in a series ascending by whole powers of T,

$$C_p = + bT + cT^2 + dT^3 + \ldots$$

then

$$E = +\frac{b}{2}T^2 + \frac{c}{3}T^3 + \frac{d}{4}T^4 + \ldots \quad . \quad (49)$$

and

$$-F = -\frac{b}{2}T^2 - \frac{c}{2 \cdot 3}T^3 - \frac{d}{3 \cdot 4}T^4 - \ldots \quad . \quad (50)$$

Usually, however, we shall make use of Einstein's or Debye's functions to represent C_p.

Using Einstein's function,

$$C_p = 3R\frac{\left(\frac{\beta\nu}{T}\right)^2 e^{\frac{\beta\nu}{T}}}{\left(e^{\frac{\beta\nu}{T}} - 1\right)^2} + kT^{3/2} \quad . \quad (51)$$

The second term on the right takes account of the correction of C_v to C_p (cf. p. 65). It follows by integration (as may be most simply verified by differentiation) that

$$E = 3R\frac{\beta\nu}{\left(e^{\frac{\beta\nu}{T}} - 1\right)} + \frac{2}{5}kT^{5/2} \quad . \quad (52)$$

IMPORTANT MATHEMATICAL FORMULÆ

and

$$-F = 3RT \log_e\left(e^{\frac{\beta\nu}{T}} - 1\right) - 3R\beta\nu - \frac{4}{15} kT^{3/2} \quad . \quad (53)$$

Tables for the more convenient manipulation of these formulæ have been calculated and are given at the end of this book. The method in which the formula due to Lindemann and myself (p. 59) is to be used is immediately clear from these; detailed tables may be found in the work of Pollitzer.

At the present time the Debye function is far the most important; according to this (cf. p. 60),

$$C_v = 3R\left[\frac{4\pi^4}{5}\left(\frac{T}{\beta\nu}\right)^3 - \frac{3\frac{\beta\nu}{T}}{e^{\frac{\beta\nu}{T}} - 1}\right.$$

$$\left. - 12\frac{\beta\nu}{T}\sum_{n=1}^{n=\infty} e^{-n\frac{\beta\nu}{T}}\left\{\frac{1}{n\frac{\beta\nu}{T}} + \frac{3}{n^2\left(\frac{\beta\nu}{T}\right)^2} + \frac{6}{n^3\left(\frac{\beta\nu}{T}\right)^3} + \frac{6}{n^4\left(\frac{\beta\nu}{T}\right)^4}\right\}\right] \quad (54)$$

Here, again, a table calculated in detail is given in the Appendix. As the treatment of the term for correcting C_v to C_p has been already given above, we now omit it: it then follows, as Debye has shown, that

$$E = \frac{9}{12} R\left(\frac{C}{C_\infty} + \frac{3x}{e^x - 1}\right)T; \quad x = \frac{\beta\nu}{T}, \quad C_\infty = 3R \quad . \quad (55)$$

Development by series gives (cf. equation **54**)

$$E = 0.75\beta\nu R\left\{\frac{77.94}{x^4} - 12\sum_{n=1}^{n=\infty} e^{-nx}\left(\frac{1}{nx} + \frac{3}{n^2 x^2} + \frac{6}{n^3 x^3} + \frac{6}{n^4 x^4}\right)\right\}$$

Integration to determine F offers no difficulties; with the aid of the well-known recurrence formula,

$$\int_x^\infty \frac{e^{-x}}{x^{n+1}} dx = \frac{1}{n}\frac{e^{-x}}{x^n} - \frac{1}{n}\int_x^\infty \frac{e^{-x}}{x^n} dx$$

we find at once (Nernst, 78 and 79)

$$F = 9R\left\{\frac{2\cdot 1646}{x^3} - \sum e^{-nx}\left(\frac{1}{n^2x} + \frac{2}{n^3x^2} + \frac{2}{n^4x^3}\right)\right\}T \quad (56)$$

If equation **(54)** is introduced into **(56)** we get

$$F = 9R\left(\frac{C}{C_\infty 36} + \frac{x}{(e^x - 1)12} + \sum \frac{e^{-nx}}{3n}\right)T, \quad . \quad (57)$$

which formula is much more convenient than **(56)** if values are calculated for $\dfrac{C}{C_\infty} = \dfrac{C}{3R}$, as has been done by Debye (*loc. cit.*, p. 803), and, in more detail, in the corresponding table in the Appendix.

My late celebrated colleague Schwarzschild kindly informed me that F may also be represented in the shorter form

$$F = 9R\left\{\frac{C}{C_\infty 36} + \frac{x}{(e^x - 1)12} - \frac{1}{3}\log_e(1 - e^{-x})\right\}T \quad (58)$$

This expression results from **(57)** by making use of the relation

$$-\log_e(1 - y) = y + \frac{y^2}{2} + \frac{y^3}{3} + \ldots$$

and putting

$$y = e^{-x}.$$

With the aid of the tables given at the end of the book the application of the above expressions to the calculation of U and A is quite simple.

Of course, one can also dispense entirely with analytical formulations and confine oneself simply to graphical representation. In this case the procedure is as follows :—

The measured values of C_p are plotted and extrapolated to the absolute zero, making use, at the lowest temperatures, of the T^3-law. Measurement of the area enclosed between the axis of temperature, the curve, and an ordinate drawn

IMPORTANT MATHEMATICAL FORMULÆ

through T gives E as a function of T. Next $\frac{E}{T^2}$ is also plotted graphically, and thus is obtained the integral $\int_0^T \frac{E}{T^2} dT$ for any required temperature, and hence the relation of F to the temperature. Frl. Miething has recently prepared tables of this kind for all solids for which sufficient measurements of C_p are available; with the aid of these the calculation for any reaction reduces to a simple addition of values taken directly from the respective tables.

As an example of such tables, short extracts from those for silver, iodine, and silver iodide are subjoined:—

Silver

T =	20	50	100	200	250	280	290
E =	1·84	47·0	243	785	1077	1257	1316
F =	0·61	19·6	154	722	1164	1449	1546

Iodine

T =	20	50	100	200	250	280	290
E =	11	124	390	973	1287	1482	1548
F =	5	78	379	1368	1989	2391	2530

Silver Iodide ½(AgI)

T =	20	50	100	200	250	280	290
E =	14	99	366	926	1234	1421	1485
F =	8	79	342	1243	1820	2198	2328

A practical application of these values is given later.

CHAPTER IX

APPLICATION OF THE HEAT THEOREM TO CONDENSED SYSTEMS *

1. The Possibility of an Experimental Proof of the Heat Theorem.—In the three previous chapters the theoretical foundations of the Heat Theorem have been explained; we shall now consider more closely its practical consequences.

The new Heat Theorem constitutes an advance which may be summarized by saying that it gives a clear relation between the curves for A and U which is not afforded by classical thermodynamics. But the suggestion arises whether, in view of the fact that its application is rendered possible only by an extrapolation to the absolute zero, or at least to the immediate neighbourhood of it, the Theorem can lead to consequences which may be realized in practice.

This question depends only on whether we are able to determine the specific heats of the substances to be considered under these circumstances. We have seen in Chapters III and IV that, at any rate in some cases, knowledge of the specific heats down to very low temperatures is to be obtained experimentally.

The extrapolation to the absolute zero itself can be made with complete safety for solids with the aid of Debye's T^3-law, once sufficiently low temperatures have been reached.

The above considerations show that a rigid experimental test of our Heat Theorem must be possible at least in some circumstances; we shall now attempt to explain the more precise requirements.

* See also Supplement, p. 265.

APPLICATION TO CONDENSED SYSTEMS 99

Crystallized bodies which undergo no allotropic transformation, or at least no rapid one, can obviously be examined at once down to near the absolute zero.

Liquids, when strongly supercooled, are as a rule frozen, and so the examination of them down to very low temperatures is rendered impossible. We know, however, a large number of exceptions, in particular the glasses; quartz glass may be mentioned as a striking example. Molten quartz, if cooled sufficiently rapidly, does not crystallize, but passes *continuously* into the condition of amorphous solid, quartz glass; the specific heat of this can be measured without any special difficulty down to temperatures as low as may be desired.

For the majority of known liquids this has not hitherto been possible, e.g. it cannot be done with water. But since the problem has been solved for a number of liquids, we cannot doubt that it can be solved in general, either by discovering means of preventing crystallization by extremely rapid cooling or by examining suitable mixtures of liquids, or adsorbed liquids, and then extrapolating to the pure state.

The matter becomes considerably more difficult in the case of *gases;* in order to be able to apply our Heat Theorem here we should have to investigate the specific heat of a gas at constant volume down to very low temperatures. According to all our experience, such an investigation is, however, quite impossible, since with sufficiently intense cooling every gas we know condenses.

With solutions the problem is similar to that with gases: here again—except in the case of solid solutions, the specific heats of which can obviously be measured down to temperatures as low as desired—we are not, in general, in a position to prevent separation, and hence a discontinuity, on intense cooling.

On these grounds I limited the application of my Heat Theorem, in my first work (1905), to pure solid or liquid condensed substances. This limitation was, however, of a purely practical, not of a fundamental nature, and it could be tolerated the more readily in that the possibility of an almost unlimited applicability resulted from a means of avoiding condensed systems (cf. Chapter X).

In the meantime it has been found by theoretical means, as we shall explain in more detail in Chapter XIV, that my Heat Theorem may be applied directly even to gases, and *a fortiori* of course to solutions. In the present chapter we shall concern ourselves with the much more simple application to condensed pure substances; this procedure corresponds also with the historical development.

2. Expansion of a Solid.—The application of the Second Law to this process is contained in the equation given on page 90:—

$$p - \frac{\partial U}{\partial v} = T\frac{\partial p}{\partial T} \quad . \quad . \quad . \quad (43)$$

Our Heat Theorem then gives at once

$$\lim \frac{\partial p}{\partial T} = 0, \text{ for } T = 0, \quad . \quad . \quad (59)$$

i.e. the pressure coefficient, like the specific heat, must vanish at low temperatures.

According to all molecular theories of solids hitherto propounded, and to all experimental evidence, the coefficient of elasticity does not change much more at low temperatures, and has therefore a finite value even at the absolute zero. Consequently the coefficient of thermal expansion must also converge towards zero at low temperatures, i.e. we must have

$$\lim \frac{\delta v}{\delta T} = 0, \text{ for } T = 0 \quad . \quad . \quad (60)$$

As Grüneisen (cf. Chapter IV) has shown, thermal expansion and specific heat must, on certain simple assumptions, be proportional to one another. The last-mentioned consequence of my Heat Theorem may therefore be most simply

APPLICATION TO CONDENSED SYSTEMS

tested on substances crystallizing in the regular system, which have a large value of βv, and hence show an early decrease in specific heat.

Actually Ch. Lindemann (55) and (58), and Röntgen * have been able to prove that there is an approximate proportionality between specific heats and thermal expansion. Since in the former case the attainment of a zero value at low temperatures may be regarded as proved, we may assume the same for thermal expansion; in any case, the papers cited, particularly Röntgen's measurements on diamond, show directly that the coefficient of thermal expansion falls off very considerably at low temperatures, as it should on our Theorem.

The circumstance that van der Waals' equation of state

$$\left(p + \frac{a}{v^2}\right)(v - b) = RT, \quad \frac{\partial p}{\partial T} = \frac{R}{v - b}$$

is in no manner of agreement with formula (59) is to be regarded as fresh evidence that this equation does not extend to the region of low temperatures.

On a point of history, it may be remarked that a connection between thermal expansion and specific heat had been pointed out in my Silliman lecture (1907, p. 122); formula (60) was derived later (1907) by Planck † and by me (38) ‡ simultaneously.

3. Melting-point.—It did not escape Berthelot that his principle is not applicable to processes like the melting of crystalline substances. As a matter of fact, the heat of fusion is always positive, i.e. heat is always evolved in the transition from the liquid to the crystalline state. At low temperatures the process does proceed as required by the principle of Berthelot, i.e. in the direction in which heat is

---

* " Ber. der Kgl. Bayr. Akad. d. Wissensch.," 1913; cf. also the numerical calculation by Grüneisen (Solvay Congress, 1913).
† " Thermodynamik," 3rd ed., p. 271.
‡ Cf. also " Theor. Chem.," p. 829.

evolved; in other words, liquids freeze at low temperatures. At high temperatures, that is, above the melting-point, the crystal liquefies, i.e. the process goes on in the direction in which heat is absorbed, being thus counter to the requirements of Berthelot's principle. Finally, at the melting-point itself both forms are in equilibrium, although of course the heat of fusion is certainly not zero.

These questions are made quite clear by our Heat Theorem. If we start from the melting-point and try to draw the U-curve towards the region of low temperatures, we obtain, from the fact that the specific heat in the liquid state is greater than that in the crystalline, the result that U decreases with falling temperatures, i.e. that U must run somewhat as shown in Fig. 13. From this the course of the A-curve is also given; whence it follows that A must cut the temperature axis. A zero value of A signifies an equilibrium between the two forms; i.e. the point of intersection is the melting-point.

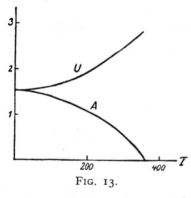

Fig. 13.

This is the place to correct a misunderstanding which has arisen in the literature, although I have repeatedly given warnings against it. The condition for equilibrium is not, strictly,
$$A = 0,$$
but
$$A + p(v' - v'') = 0,$$
where v' and v'' are the specific volumes of the two modifications. But at moderate pressures the second expression appears as a vanishingly small correction term, neglect of

APPLICATION TO CONDENSED SYSTEMS

which is justified and usual. For condensed systems atmospheric pressure is sufficiently low, and since the determination of thermal quantities has been made almost exclusively at atmospheric pressure, we shall make use, in what follows, of the simplified expression. More exactly, we shall assume henceforth that all thermal measurements have been reduced to the value $p = 0$. The method of taking account of the influence of pressure is supplied by classical thermodynamics, and need not be detailed here.

As regards the quantitative test of the Heat Theorem in the present case there must be measured, in addition to the heat of fusion, the specific heats of the crystallized substance and of the supercooled liquid down to low temperatures, if possible into the region of the T^3-law in both cases. The former series will not now in general offer any particular difficulty, but the examination of the liquid form is hindered by the fact that supercooling down to temperatures as low as desired is possible only in the rarest cases.

Quartz would be a good example, since the melt can be supercooled to any extent in the form of quartz glass; the specific heats have here been measured for both forms down to low temperatures, but the heat of fusion has not yet been determined.

It is known from the investigations of Tammann that benzophenone and betol are capable of extensive supercooling; but difficulties, which have not yet been fully explained (Koref, 60), have been encountered in making accurate determinations, for the values obtained for the specific heats show unusually large variations. Examination of the supercooled substances was not possible in the vacuum calorimeter; with the copper calorimeter the following numbers were obtained (60):—

Benzophenone.		Betol (graphical interpolation)	
T.	$\dfrac{dU}{dT}$	T	$\dfrac{dU}{dT}$
137	0·1526 − 0·1514 = 0·0012	130	0·148 − 0·144 = 0·004
295	0·3825 − 0·3051 = 0·0774	240	0·256 − 0·2205 = 0·0355
		320	0·362 − 0·295 = 0·067

The above values are not sufficiently reliable for an accurate determination of the melting-point, as already explained ; nor do they extend over a wide enough interval of temperature. They are sufficient, however, to give a rough picture of the A- and of the U-curves, and a graphical representation will at once show that the position of the melting-point may be derived, at least approximately, from the heats of fusion * and the above values of specific heat.

The question arises whether, in this apparently simple case, there is not a specially simple formula for the shape of the A- and U-curves.

In this connection I was led to the simple formulation,

$$U = U_0 + \beta T^2, \qquad A = U_0 - \beta T^2, \qquad (61)$$

according to which the two curves would be symmetrically disposed one on each side of the common tangent ; this formulation results from limiting the development of the series in equations **(36)** and **(37)** of p. 81 to the first term. This would mean that the difference between the specific heats in the liquid and crystallized states would be given by the relation

$$c - c' = 2\beta T.$$

For the melting-point (A = 0) this leads to

$$T_0^2 = \frac{U_0}{\beta}.$$

* Measured by Tammann, " Zeitschr. physik. Chem.," **39**, 63 (1889).

APPLICATION TO CONDENSED SYSTEMS

If we denote the specific heats by c_1^0 and c_2^0, and the heat of transformation at the melting-point T_0 by U_0, it is easy to derive the further relation

$$T_0 = \frac{U_0}{c_2^0 - c_1^0}.$$

Thus this equation holds only on the assumption that the difference in specific heat between the solid substance and the supercooled liquid increases proportionally with the absolute temperature. In many actual cases this appears to be approximately true: thus Tammann * discovered the above relation purely empirically; e.g. for naphthalene

$$U_0 = 34\cdot 7, \ c_1^0 = 0\cdot 332, \ c_2^0 = 0\cdot 442,$$

and hence

$$T_0 = \frac{34\cdot 7}{0\cdot 11} = 315 \text{ (instead of 353)}.$$

Water, as is well known, exhibits great anomalies in regard to specific heat in the liquid state, as the result, apparently, of strong association; hence the development for U should not be stopped at the second term, and consequently the above relation is not even approximately correct, as it gives

$$T_0 = \frac{80}{1\cdot 00 - 0\cdot 51} = 163 \text{ (instead of 273)}.$$

In most of its relationships water is included with those substances which have an abnormal behaviour: Tammann's relation should therefore be limited to the so-called normal † liquids, for which it appears to hold fairly approximately.

It should, however, be expressly noted that this can only be an approximation formula, since at low temperatures equations **(61)** are to be replaced by formulæ **(41)**.

$$U = U_0 + \delta T^4, \quad A = U_0 - \frac{\delta}{3}T^4.$$

* " Kristallisieren und Schmelzen," pp. 42 *et seq.* (Leipzig, 1903).
† Cf. " Theor. Chem.," p. 316.

4. Transition Point.

—Exactly the same considerations and equations hold, of course, for allotropic change, the modification stable above the transformation point taking the place of the liquid. On the experimental side, however, we have in this case the great advantage that the transformation frequently takes place so slowly that we can investigate both modifications without any particular difficulty down to the region of the lowest temperatures.

It is true that, on account of the smallness of the heats of transformation, the difference in specific heat between the two modifications is not large and that the form of the A-curve can therefore be established with only limited accuracy. All available observations on sulphur, which is the only example of this kind on which detailed investigation has been made, show that, as required by my Heat Theorem, there is a contact, a very close contact, between the curves for A and U at low temperatures.

We are dealing here, of course, with the transformation of monoclinic into rhombic sulphur. Let us first make the simple hypothesis that for the transformation of 1 gramme of sulphur

$$U = U_0 + \beta T^2,$$

i.e. let us assume that since, according to the First Law,

$$\frac{dU}{dT} = 2\beta T = c_2 - c_1$$

(c_2 and c_1 being the specific heats of the two modifications), the difference between the specific heats of the two forms increases as the absolute temperature: from all the available observations we deduce the following values for U_0 and β:—

$$U = 1 \cdot 57 + 1 \cdot 15 \times 10^{-5} T^2.$$

The above simple hypothesis accords, to quite a good degree of approximation, with all measurements hitherto made; thus it reproduces well the available values of U:—

APPLICATION TO CONDENSED SYSTEMS

T.	U (obs.).	U (calc.).	Observer.
273	2·40	2·43	Broensted.
368	3·19	3·13	Tammann.

Further, it gives with satisfactory approximation the values of A found by Broensted from determinations of the solubility of the two modifications:—

$$A = 1\cdot 57 - 1\cdot 15 \times 10^{-5}\, T^2.$$

T.	A (obs.).	A (calc.).
273	0·72	0·71
288·5	0·64	0·61
291·6	0·63	0·59
298·3	0·57	0·55

Lastly, we find, from $A = 0$,

$$T_0 = \sqrt{\left(\frac{1\cdot 57}{1\cdot 15 \times 10^{-5}}\right)} = 369\cdot 5 \text{ (instead of } 273\cdot 1 + 95\cdot 4 = 368\cdot 5\text{)}.$$

The truth of the above formulæ for U and A can be particularly closely tested on the specific heats. When I proposed these equations there were only a few isolated measurements available at high temperatures; Koref and I checked and extended them, particularly in the region of low temperatures.

We have found, above, the relation

$$\frac{dU}{dT} = c_2 - c_1 = 2\cdot 30 \times 10^{-5}\, T,$$

which is tested in the following table:—

TABLE VI

T.	$\frac{dU}{dT}$	$2\cdot30\,T \times 10^{-5}$	Observer.
83	0·0854 − 0·0843 = 0·0011	0·0019	Nernst.
93	0·0925 − 0·0915 = 0·0010	0·0021	,,
138	0·1185 − 0·1131 = 0·0054	0·0032	Koref.
198	0·1529 − 0·1473 = 0·0056	0·0046	Nernst.
235	0·1612 − 0·1537 = 0·0075	0·0054	Koref.
290	0·1774 − 0·1720 = 0·0054	0·0067	Wigand.
293	0·1794 − 0·1705 = 0·0089	0·0067	Koref.
299	0·1809 − 0·1727 = 0·0082	0·0069	Wigand.
329	0·1844 − 0·1764 = 0·0080	0·0076	Regnault.

The variation of specific heat is thus in fact such as is required by the above formulæ and hence (qualitatively) by the Heat Theorem. Moreover, there are no systematic deviations which can be recognized, and I remarked on the above table (" Theor. Chem.," 7th German edition, p. 736) in 1911 thus :—

" Errors of observation will not account for the fact that the difference * in question is a little smaller at low temperatures, and at high temperatures a little greater, than corresponds with the above formula ; but the discrepancies are too small to affect the accuracy of the above equations appreciably. The experiments have certainly given good confirmation of the variation of the specific heats demanded by the foregoing formulæ."

In 1912 appeared the paper of Debye mentioned on page 59 ; according to this the above hypothesis, attractive though it is in view of its simplicity,

$$U = U_0 + \beta T^2, \quad A = U_0 - \beta T^2, \qquad (61)$$

can no longer be maintained, since in the region of the lowest temperatures the equations

$$U = U_0 + \delta T^4, \quad A = U_0 - \frac{\delta}{3} T^4 \qquad (41)$$

---

* $c_2 - c_1 = 2\cdot30\,T \times 10^{-5}$.

APPLICATION TO CONDENSED SYSTEMS

must hold for all condensed systems. Thus $\frac{dU}{dT}$ must fall off more rapidly at low temperatures than is required by equations (**61**); this is manifest in Table VI above.

Fig. 13 (p. 102) shows the shape of the A- and U-curves according to the formulæ (**61**); Fig. 14 includes an illustration of the two curves obtained graphically by means of the device described on page 84, in which, therefore,

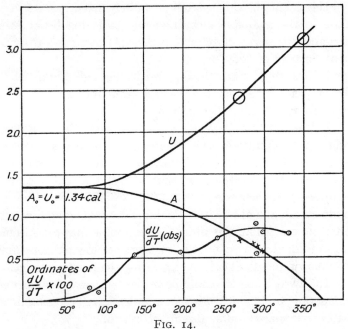

Fig. 14.

without making use of an equation, the values of $\frac{dU}{dT}$ obtained in Table VI are solely employed in drawing the U-curve. The curve was drawn very carefully on a large scale by Herr Drägert, and Fig. 14 was prepared with a view to photographic reproduction on a smaller scale.

At the bottom is the $\frac{dU}{dT}$ curve, which is used in the drawing of the U-curve; it is taken from Table VI, but in view of the rather large errors of observation it cannot of course be very smooth ; the large number of measurements nevertheless diminishes the uncertainty. The value of Wigand falls off the curve, and has therefore not been taken into consideration.

The U-curve has also been so placed as to fit, as well as possible, the two available measurements of page 107 : these are distinguished by circles.

The A-curve, which is now definitely fixed, passes very close to the available measurements (indicated by crosses, cf. table, p. 107) ; it intersects the axis of temperature at 370° (instead of 368·5°).

It has thus been possible, with the aid of our Heat Theorem, to construct the A-curve from the thermal data, so as to agree almost perfectly with the facts.

5. The Binding of Water of Crystallization or of Hydration.—The affinity of this reaction can of course be calculated from the vapour pressure of the salt containing the water of crystallization and that of water at the required temperature ; the measurement of the specific heats of the anhydrous salt and of the salt containing water of crystallization offers no particular difficulties. The trouble that the specific heat of liquid water cannot be measured at low temperatures may be simply obviated by calculating the reaction to ice.

The Heat Theorem could be tested by many examples of this sort ; we shall deal with some which have been examined with particular care (cf. also Schottky, 16).

Copper Sulphate.—The reaction

$$CuSO_4 + H_2O = CuSO_4 . H_2O$$

has been examined by A. Siggel (81) : he measured the heat of hydration with liquid water, the dissociation pressures at high temperatures, and the specific heats of the two salts and of ice down to very low temperatures.

The dissociation pressure π could then be calculated for the ordinary zero of temperature by means of the Second

Law: calling p the vapour pressure of ice at this temperature, it was found that

$$A = RT \log_e \frac{p}{\pi} = 4415 \text{ cals.}$$

On the other hand, from the heat of this reaction at the same temperature (4910 cals.) and the specific heats, the new Theorem gave

$$A = 4475 \text{ cals.},$$

in satisfactory agreement with the above value (Fig. 15). Finally, for the absolute zero, calculation gives

$A_0 = U_0 = 4680$ cals.

Calcium Hydroxide.—The reaction

$CaO + H_2O = Ca(OH)_2$

has been examined in an analogous manner by Drägert (102); here we are concerned with a very strong or, as it is often expressed, with a more chemical combination of the water.

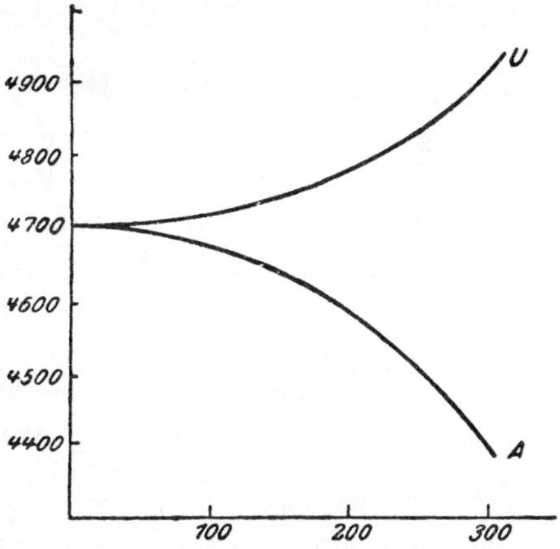

$CuSO_4 + H_2O = CuSO_4 . O_2H$
FIG. 15.

The heat of hydration was measured at 15° (15,175 cals.); calculated to ice at T = 273, this gives

$$U_{273} = 13{,}630 \text{ cals.}$$

The U-curve can now be drawn from the specific heats, or more correctly speaking, from the molecular heats; from this the A-curve can be found by means of the procedure described on page 84, as is shown in Fig. 16. Hence it follows that

$$A_{273} = 13{,}330.$$

Alternatively, the dissociation pressures of calcium hydroxide have been measured with great accuracy over the interval from 300° to 444°, and calculated to 0° (on the ordinary scale) by means of the Second Law. This is possible since the heat of dissociation and the molecular heats concerned (water vapour, $Ca(OH)_2$ and CaO) are well known. Thus for 0° C

$$\pi = 6 \cdot 0 \times 10^{-11} \text{ mm.},$$

and hence, in the same manner as above,

$$A_{273} = RT \log_e \frac{p}{\pi} = 13{,}540 \text{ cals.}$$

The agreement between the values calculated from

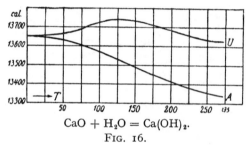

$CaO + H_2O = Ca(OH)_2$.
FIG. 16.

thermal data and from the dissociation pressures lies within the error of experiment.

Potassium Ferrocyanide.—In all the examples hitherto considered, it may be seen that U increases at first with the temperature, while A decreases: this appears to be the most frequent case. In the two cases last considered, the A-curve would not intersect the temperature axis until temperatures had been reached so high that ice had long since ceased to exist; in practice, therefore, there is no transformation point.

This form of curve will always be given when the molecular heat of the water of crystallization is less than that of ice; an example of the reverse case is given by potassium ferrocyanide (36), which crystallizes with three molecules of

APPLICATION TO CONDENSED SYSTEMS

water. The accompanying diagram (Fig. 17) shows the peculiar energy relationships. From this it is seen that at $T = 160$, U becomes 0, and at higher temperatures U is negative, while A remains positive and in fact increases. Ice and water-free potassium ferrocyanide can therefore combine to form the hydrate, although the heat of combination is negative, thus totally contradicting the Principle of Berthelot. This is, however, in full agreement with the consequences of the new Heat Theorem, which refers this behaviour to the exceptionally high specific heat of the water of crystallization in potassium ferrocyanide. The

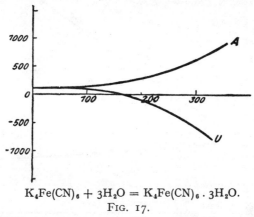

$$K_4Fe(CN)_6 + 3H_2O = K_4Fe(CN)_6 \cdot 3H_2O.$$
FIG. 17.

curves in Fig. 17 are somewhat extrapolated beyond the melting-point of ice; they are drawn in accordance with measurements, some by Schottky and some by myself.

6. Affinity between Silver and Iodine.—When I compared the heat evolution and affinity in this case, greater differences appeared at first than were to be expected, on our Theorem, from the variation of the specific heats. I therefore set Herr Ulrich Fischer (74) the detailed examination of the case; he showed conclusively that Thomsen's value for the heat of formation (13,800 cals.) was very considerably in error. From the temperature coefficient of the E.M.F. of the silver-iodine electrode, Fischer found 15,170 and, by two totally different thermochemical methods, 14,820 and 14,980, giving a mean of 14,990, whereas my Heat Theorem gave 15,080. This is therefore one of the already fairly numerous cases where this Theorem has led to the discovery of errors of measurement.

Koref and Braune (*vide infra*) later confirmed Fischer's calorimetric measurement, finding for the heat of formation of silver iodide the value 15,100.

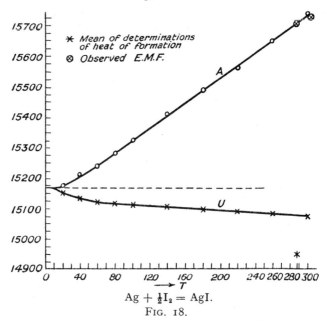

$Ag + \tfrac{1}{2}I_2 = AgI$.

Fig. 18.

The curve given above (Fig. 18), which is taken from Fischer's paper, shows the course of the two energy curves; up to about 10° abs. there is practically perfect contact, and at about 20° the difference is only a little more than 1 pro mille. At first, as required by equations **(41)**, U departs from the dotted common prolongation of the line of contact more rapidly than does A. But then, as a result of the law of Kopp and Neumann coming more or less into operation, U shows a tendency to become independent of temperature,

APPLICATION TO CONDENSED SYSTEMS

whereas, in agreement with the fundamental equation **(1)**, A varies almost linearly with the temperature. This behaviour is apparently typical.

The above numbers were calculated by Fischer by assuming the formula proposed by Lindemann and myself. When Debye's paper on specific heats appeared I went through the corresponding calculation on the assumption of Debye's function for the specific heats; as was to be expected, practically identical values were obtained.

Nor does the fact that, according to Gunther's experiments (cf. p. 53), somewhat different values should be used for the atomic heat of iodine, make much difference. By using the new table calculated by Frl. Miething (p. 97), the result is as follows :—

Following Fischer, we start with a well-established value for A, obtained by measurement of the potential, e.g. at $T = 288$; we thus get

$$\begin{aligned} U_0 &= A_{288} + F_{Ag} + F_I - F_{AgI} \\ &= 15{,}700 + 1520 + 2502 - 4605 \\ &= 15{,}117. \end{aligned}$$

The heat of formation at $T = 288$ is then calculated as

$$\begin{aligned} U_{288} &= 15{,}117 + E_{Ag} + E_I - E_{AgI} \\ &= 15{,}117 + 1309 + 1531 - 2944 \\ &= 15{,}014. \end{aligned}$$

This example will show also how much the calculation is simplified by the use of the tables mentioned.

A brief reference may be made, in conclusion, to some interesting points in the work of Braune and Koref. Ordinary (pure) AgI is usually a mixture of the amorphous and crystalline substances, and this was the case both for the iodide used by U. Fischer and that first employed by me. In the potential measurements, which last a long time, the amorphous form is changed into the crystalline, so that the heat of formation obtained by Fischer from the temperature coefficient of the silver-iodine electrode refers to the crystalline modification ; this is also true both for the later measurements by Koref and Braune and for the measurements of specific heat made by Schwers and myself (cf. p. 52). Only the following numbers are therefore strictly comparable :—

Determination of U by Fischer by electrochemical means . . . . 15,170 cals.
Determination of U by Koref and Braune thermochemically . . . . 15,100 ,,
Calculation of U from A, making use of the Heat Theorem and of the specific heat of AgI as measured in paper (95) . 15,014 ,,

The most probable value should be 15,050. The value 14,990 obtained thermochemically by U. Fischer is somewhat smaller, corresponding with the fact that the transition from amorphous to crystalline AgI develops a little heat.

7. The Clark Cell.—This galvanic combination measures the affinity of the reaction

$$Zn + HgSO_4 + 7H_2O = ZnSO_4 . 7H_2O + Hg.$$

In ordinary circumstances the system concerned is not a condensed one, but there are also processes going on in solution. If, however, we cool down to $-7°$, ice appears as a solid phase, and then the above reaction occurs between pure substances simply. The heat of this reaction is well known and, in addition, all the reactants have been measured to a sufficiently low temperature (that of boiling hydrogen). The heat change is first calculated at $T = 234$, at which temperature the mercury is solid and the heat therefore changes by 2×555 cals. (heat of fusion of mercury). From this point the U-curve can be drawn down to the absolute zero, and the A-curve calculated with the aid of the Heat Theorem. This calculation, for which we have to thank Pollitzer (46), gives for $T = 266$ an E.M.F. of 1·456 volts, which is identical with the observed value. Consequently the potential, which is of course also a measure of the chemical affinity, can be calculated for the cryohydric point without making any assumptions.

It is, to be sure, accidental that this concordance is so close, for in this case, on account of the large number of atoms taking part in the reaction, the influence of the specific heats is relatively large, and the error of experiment may therefore easily cause an error of 0·01 volts.

The accompanying curves (Fig. 19) give a good idea of the conditions. From low temperatures up to the melting-point of mercury there is hardly any appreciable difference between A and U ; at this point, for example,

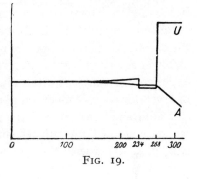

FIG. 19.

$$U = 67,400, \; A = 67,000 \text{ cals.}$$

Owing to the melting of mercury U now falls a little below A, and A must therefore bend a little upwards. At $T = 266$, since the heat of dilution can be neglected, U increases suddenly by the very considerable heat of fusion of ice and, as U is now much greater than A, $\frac{dA}{dT}$ becomes relatively strongly negative, which corresponds with experience.

It may be seen that, by combination of the new Heat Theorem with the two existing laws of thermodynamics, the energy relationships in this case can be completely explained. The contact between the A- and U-curves required by the new Heat Theorem can here be noted as practically complete up to relatively high temperatures; this is a common occurrence.

In particular, the relatively large temperature coefficient of the electromotive force has found a complete thermodynamical explanation; it is the conversion of 7 mols of ice into liquid water which mixes with the solution, which is the essential reason for the shift of the U-curve and the consequent considerable value of $\frac{dA}{dT}$.

No analogous investigation of the Weston cell has yet been made. There are certain difficulties in the way because, on account of transformation of the salt $CdSO_4, \tfrac{8}{3}H_2O$, the cryohydric point cannot be reached directly. The appreciably smaller temperature coefficient of this cell is partially explained, of course, by the much smaller amount of water of crystallization ($\tfrac{8}{3}$ mols compared with 7 in the Clark cell); but additional reasons are the great solubility of cadmium sulphate at low temperatures, and the fact that the cadmium salt, in contrast with the zinc salt, dissolves with evolution of heat.

8. Investigations of Braune and Koref.—We are indebted to Braune and Koref (77, and particularly 98b) for an extremely careful and painstaking test of the application of the Heat Theorem to condensed systems. The test was made on a number of galvanic cells in exactly the same manner as that already employed by U. Fischer. In all the cells U was found to have nearly the same value, whether determined from the temperature coefficient of the potential or by direct thermochemical means; this provides a guarantee that the cells under observation were actually controlled by the process assumed to be supplying the current, though as a matter of fact there was hardly any doubt of this in the cases examined.

In the following table are given the values for $U_0 = A_0$ per electrochemical gramme-equivalent; these were calculated, by means of the specific heats, from the affinity A at room temperatures, deduced from the potential; U_1 is the value obtained from $\frac{dA}{dT}$, U_2 that found thermochemically, and U that deduced from the Heat Theorem:—

APPLICATION TO CONDENSED SYSTEMS

The values in the last column but one, calculated by means of the Heat Theorem, agree extremely well with the mean of U_1 and U_2 (last column but two): the differences, sometimes positive and sometimes negative, are all within the errors of observation which affect the values of U_1 and U_2, and also of the specific heats, and may easily amount to 200 cals. by accumulation. The mean value of U, U_1, and U_2 should give us the heat evolution in the respective reactions with an accuracy hitherto unknown in such cases.

9. Summary.—We shall not need to deal with any further examples: in none of the cases hitherto examined has there been even the slightest occasion to doubt the correctness of the Heat Theorem. On the contrary, there have been repeated instances where, when serious discrepancies at first arose, unexpectedly large errors were discovered in the experimental determination of one of the magnitudes concerned.

We have drawn the U and A diagrams, above, for a number of chemical and physico-chemical reactions; the question arises

TABLE VII

Reaction.	$U_0 = A_0$.	A.	U_1.	U_2.	Mean.	U.	Difference.
$Pb + I_2 = PbI_2$	41,874	41,220	41,960	41,850	41,905	42,034	+ 129
$Ag + I = AgI$	15,228	15,715	15,169	15,100	15,135	15,014 §	+ 121
$Pb + 2AgCl = PbCl_2 + Ag$	24,262	22,540	24,748	24,880	24,814	24,732	— 82
$Hg + AgCl = HgCl + Ag$	332	—530 *	1,382	1,427	1,405	1,270	— 135
$Pb + 2HgCl = PbCl_2 + 2Hg$	23,597	24,020 *	21,940	22,070	22,005 †	22,159	+ 154

* Reduced to T = 234 (solid mercury).
† Given wrongly in the original by an obvious printer's error.
§ Cf. above, p. 115.

—are there any further special regularities in the course of the two thermodynamical functions? By special regularities are to be understood such as are not fixed either by classical thermodynamics or by our Heat Theorem.

As far as the available material is concerned, the answer must be that there are apparently none; every case thermodynamically possible appears to occur in nature.

It might, for example, have been conceivable that Fig. 14 or 15 gave us the typical form, i.e. that, if we regard the process in question as taking place in the direction in which

$$U_0 = A_0 > 0,$$

U increases steadily with the temperature and A correspondingly decreases.

Such is not the case, however, as is shown by Fig. 18 and many other well-investigated instances. If there had been such a rule we should have had the advantage of being able to state the course of the reaction (i.e. the sign of A) for low temperatures, and with some degree of probability for higher temperatures also, from a measurement of the specific heats of the substances concerned at any one temperature.

In practically all cases hitherto observed the rule holds that the difference U — A increases in absolute amount with rising temperature; but here again there are exceptions (cf. Fig. 16).

From the theoretical aspect the following may be noted regarding this question. It is required to find general relationships for the temperature function

$$U = f(T) \ ;$$

this depends purely on whether there are such relationships between the specific heats of the substances under consideration. This is not probable, and it is hardly a thermodynamical problem pure and simple, because the specific heats depend on the special nature of the materials.

CHAPTER X

APPLICATION OF THE HEAT THEOREM TO SYSTEMS CONTAINING A GASEOUS PHASE

1. Statement of the Problem.—In reactions in which gases take part, where, therefore, there is a gaseous phase in the equilibrium, it is necessary for an immediate application of the Heat Theorem that we should be in a position to determine the specific heats down to the absolute zero, even for the gases concerned.

The same difficulty arises here, but in a greater degree, as that which we encountered on page 99 in the consideration of processes in which liquids took part—how can it be really possible to examine a quantity of gas for heat capacity at constant volume down to low temperatures, without the gas condensing?

We shall see in Chapter XIV that it appears possible to overcome these difficulties, to some extent at least, partly by theoretical and partly by experimental means, and that there can hardly be any doubt at the present time that a complete solution of the problem is attainable.

This reasoning is affected, however, by various more or less hypothetical assumptions; up till now we have been moving, we may say, on absolutely safe ground, and we shall not leave it if we content ourselves, in this and the following chapter, with an indirect application of the Heat Theorem. This is possible if we apply the Theorem to condensed systems and operate, as far as the gaseous phase is concerned, solely with classical thermodynamics. A simple example may at once show us how we have to proceed.

We have seen above how the affinity of the condensed reaction

$$2Ag + I_2 = 2AgI,$$

may be calculated from purely thermal data; also, this value is

$$A = RT \log_e \frac{p}{\pi}.$$

If p, the vapour pressure of solid iodine, is known at any given temperature, we can hence calculate π, the dissociation tension of silver iodide at this temperature. We know the equilibrium between solid silver, silver iodide, and gaseous iodine for all temperatures if the vapour-pressure curve of iodine is given; according to classical thermodynamics this is the case if, in addition to the specific heat of solid iodine already required for the calculation of A, we know that of iodine vapour as well, and also either the heat of evaporation and the vapour pressure at one temperature, or the vapour pressures at two temperatures.

This example shows us then that, in order to get an idea of the behaviour of a gas, we must first consider the gas in question as taking part in the equilibrium as a " precipitate " (i.e. in a suitably condensed form). There is then nothing to hinder the application of the Heat Theorem and, with the aid of the vapour pressure, the transition to the gaseous state is given.

2. Vapour-pressure Curve.—In view of the fundamental importance which the vapour-pressure has now assumed for our problem—the calculation of chemical equilibria and affinities—let us summarize what we may learn from classical thermodynamics towards calculating the vapour-pressure curve, making use also of the information we have lately obtained concerning specific heats.

If we limit ourselves to *small* vapour pressures, we have

$$\lambda = RT^2 \frac{d \log_e p}{dT} \quad . \quad . \quad . \quad (62)$$

$$U = RT^2 \frac{d \log_e \xi}{dT} \quad . \quad . \quad . \quad (63)$$

APPLICATION TO A GASEOUS PHASE

Here λ is the ordinary heat of vaporization per mol, U the heat development associated with the condensation without doing external work; p and ξ are the pressure and concentration of the saturated vapour. The two equations are fundamentally identical; they result from the Clausius-Clapeyron equation by introducing the laws of the ideal gaseous state and limiting the application to small pressures.

Now let λ be given as a function of temperature down to the absolute zero; to fix our ideas suppose it to be in the form of the series

$$\lambda = \lambda_0 + a_0 T + \beta_0 T^2 + \gamma_0 T^3 + \ldots \qquad (64)$$

Then integration of (62) gives

$$\log_e p = -\frac{\lambda_0}{RT} + \frac{a_0}{R} \log_e T + \frac{\beta_0}{R} T + \frac{\gamma_0}{2R} T^2 + \ldots + i \quad (65)$$

where i denotes the constant of integration. Substituting

$$p = \xi RT \qquad (66)$$

this gives

$$\log_e \xi = -\frac{\lambda_0}{RT} + \left(\frac{a_0}{R} - 1\right) \log_e T + \frac{\beta_0}{R} T$$
$$+ \frac{\gamma_0}{2R} T^2 + \ldots + i' \quad (67)$$

where we have put

$$i' = i - \log_e R \qquad (68)$$

Equation **(67)** may also be obtained by direct integration of **(63)**.

It is to be particularly noted that, in order to determine the constant i (or i'), we must know λ as a function of temperature down to very low temperatures; we cannot determine i even approximately from a vapour-pressure equation, valid though it may be for a large interval, if it does not hold down to the lowest temperatures.* To mention an

---

* The calculation of a vapour-pressure curve, according to formula **(67)**, over a wide range of temperature was first rendered possible by our measurements of specific heat; the vapour-pressure curve of ice down to very low temperature is given by Nernst (39).

example, we may use the equation
$$\lambda = \lambda_0 - 1\cdot 68 T,$$
which holds very accurately for the heat of vaporization of liquid mercury over the interval $-34°$ to $+260°$: for in this interval liquid mercury has the constant atomic heat 6·64 and gaseous mercury (at constant pressure) 4·96, and $6\cdot 64 - 4\cdot 96 = 1\cdot 68$. Hence in this interval, but only here,

$$\log_e p = -\frac{\lambda_0}{RT} - 0\cdot 847 \log_e T + B,$$

which equation, as a matter of fact, reproduces the measurements with perfect accuracy; the numerical value of B, however, is not by any means to be identified with i (cf. in particular Chapter XIII).

In order to avoid mistakes with equations which, like the foregoing, are expressly intended to be valid only for a limited interval of temperature, it is desirable to characterize them as such by writing them in some such way as

$$\log_e p = -\frac{\lambda_0}{RT} - 0\cdot 847 \log_e T + B \Big]_{-34°}^{+260°}$$

Equation **(65)** assumes a very simple form for very low temperatures and a monatomic vapour; for here the specific heat of the condensed phase may be neglected, while the molecular heat of the vapour may be taken as

$$C_p = \tfrac{5}{2} R\,;$$

thus under these conditions *

$$\log_e p = -\frac{\lambda_0}{RT} + \frac{5}{2} \log_e T + i \qquad . \qquad . \quad \textbf{(69)}$$

We shall return to this very simple and important equation of state in Chapter XIII.

* Nernst, "Solvay Congress, 1911," p. 233 (German edition, "Abhandl. d. Bunsenges.").

APPLICATION TO A GASEOUS PHASE

3. Chemical Equilibrium in Homogeneous Gaseous Systems.—Let us consider the equilibrium of the reaction

$$n_1 A_1 + n_2 A_2 + \ldots = n_1' A_1' + n_2' A_2' + \ldots$$

If the concentrations of the molecular species $A_1, A_2, \ldots$ are $c_1, c_2, \ldots$, and those of the molecular species $A_1', A_2', \ldots$ are $c_1', c_2', \ldots$, the law of mass action gives

$$K = \frac{c_1^{n_1} c_2^{n_2} \ldots}{c_1'^{n_1'} c_2'^{n_2'} \ldots} \qquad . \qquad . \qquad (70)$$

and the equation of the reaction isochore is

$$U = RT^2 \frac{d \log_e K}{dT} \qquad . \qquad . \qquad (71)$$

If the change of energy in the reaction is given by the expression

$$U = U_0 + \alpha T + \beta T^2 + \gamma T^3 + \ldots, \qquad (72)$$

integration gives

$$\log_e K = -\frac{U_0}{RT} + \frac{\alpha}{R} \log_e T + \frac{\beta}{R} T + \frac{\gamma}{2R} T^2 + \ldots + I \quad (73)$$

In order to be able to apply our Heat Theorem, let us consider the equilibrium at temperatures so low that all the reacting molecular species are possible as condensates; we can then deal with the same reaction simply between solid and liquid (pure) substances, when it will take place according to the scheme

$$n_1 a_1 + n_2 a_2 + \ldots = n_1' a_1' + n_2' a_2' + \ldots \qquad (74)$$

where the values of a now refer to the solid state.

For the reaction $a = A$ (sublimation process) the law of mass action gives, of course,

$$K = \xi,$$

and we then find again by this method the vapour-pressure formula **(67)**—

$$\log_e \xi = -\frac{\lambda_0}{RT} + \left(\frac{a_0}{R} - 1\right) \log_e T + \frac{\beta_0}{R} T + \frac{\gamma_0}{2R} T^2 + \ldots + i',$$

where it is assumed that equation **(64)** holds for the heat of evaporation.

By a simple calculation * we find for the affinity A of the reaction **(74)**,

$$A = - RT(\log_e K - \Sigma n \log_e \xi) \quad . \quad . \quad (75)$$

where $\Sigma n \log_e \xi$ signifies the summation

$$n_1 \log_e \xi_1 + n_2 \log_e \xi_2 + \ldots - n_1' \log_e \xi_1' - n_2' \log_e \xi_2' \ldots \quad (76)$$

Now the Heat Theorem states that, for a reaction on the scheme of equation **(74)**, the coefficients of the terms T and T $\log_e$ T in equation **(75)** must vanish; if we combine this equation with **(73)** and **(67)** we see at once that the term of the series which contains T as a factor † is

$$RT(I - \Sigma n i'),$$

and thus

$$I = \Sigma n i' \quad . \quad . \quad . \quad (77)$$

This result,‡ which is certainly a remarkable one, has the following meaning. The equation **(73)** contains, on the right-hand side, in addition to thermal quantities, only the constant of integration I. This is now referred to a sum of integration constants which can be determined once for all for each molecular species, most directly from the vapour-pressure curves of the substances concerned in the liquid or solid state. At the same time, the above calculations show that the constant of integration i' (and hence also i) is independent of the nature of the condensation product, e.g. the values for ice and liquid water are identical. Neither of the two older Laws has anything to say on this subject.

4. Heterogeneous Equilibrium.—We shall now assume that in a homogeneous gaseous system, for which, following the method developed in the previous section, the relation

---

* Cf. e.g. " Theor. Chem.," p. 744.

† The condition that the factor of T $\log_e$ T should also disappear is fulfilled owing to the additivity of the atomic heats of solids at very low temperatures, and so tells us nothing more.

‡ See Eucken and Fried, " Zeits. f. Physik," **29**, 36, 1924, and **32**, 150, 1925, for some objections to its validity (Tr.).

$$\log_e K = -\frac{U_0}{RT} + \frac{a}{R}\log_e T + \frac{\beta}{R}T$$
$$+ \frac{\gamma}{2R}T^2 + \ldots + \Sigma ni' \quad (78)$$

holds, there is one molecular species participating as a condensed phase. We then subtract from the above equation the expression

$$n \log_e \xi = n\left\{-\frac{\lambda_0}{RT} + \left(\frac{a_0}{R} - 1\right)\log_e T + \frac{\beta_0}{R}T\right.$$
$$\left. + \frac{\gamma_0}{2R}T^2 + \ldots + i'\right\}$$

which relates to the condensate in question. We thus make the equilibrium constant K have the value which corresponds with the heterogeneous equilibrium in question according to the law of chemical mass action. The value of i' pertaining to this molecular species consequently disappears from the last term; the remaining terms, which contain purely thermal quantities, assume, similarly, the values which pertain to the heterogeneous system.

If there are several molecular species taking part as condensates, the above operation is to be repeated as often as required, and we thus arrive, finally, at the simple result that equation **(77)** is also applicable to heterogeneous systems in which there are any number of solids co-existing with a gaseous phase. The thermal quantities then relate to the heat change of the reaction concerned and, in forming the quantity K, only those molecular species are to be considered which do not at the same time co-exist as condensates. Consequently, only the i' values for these species have to be summed in order to obtain the integration constant I.

The practical application of the above formulæ requires that the specific heat of the gases should be known in the low temperature range where our experimental data are still incomplete (cf. Chapter V); we shall return to this question in Chapters XIII and XIV.

CHAPTER XI

A THERMODYNAMICAL APPROXIMATION FORMULA

1. General.—When I put forward my Heat Theorem, our information on the subject of the specific heats of solids at low temperatures was so scanty that it was quite impossible to make a direct test such as we have described in Chapter IX for condensed systems. In order to assure myself that I was on the right track, I attempted to make a test, which should be at least approximate, of the formula

$$I = \Sigma n i' \quad . \quad . \quad . \quad . \quad (77)$$

If this proved favourable, the Heat Theorem was thus proved at the same time for chemical equilibria in general, for from the foregoing formula the rule

$$\lim \frac{dA}{dT} = 0 \text{ (for T = 0)}$$

for condensed systems may also be derived by reversing the reasoning.

I recognized, of course, quite clearly that this could only be an approximate test. But if the regularities resulting from formula **(77)** were found to be actually confirmed in nature, this would constitute good support. As my colleague Planck once said to me in conversation, one cannot but assume that a law so general as this, that the entropy constant of matter is negligibly small at the absolute zero, is either as exactly true as the other Laws of Thermodynamics, or else is sometimes far removed from the truth. In spite of much search, no case has yet been found in which this law is totally contradicted : nay, rather, it has eventually proved to hold even in cases in which apparently reliable measurements were at first in conflict with it. Strong support has thus been obtained for this Theorem.

Now the equation

$$I = \Sigma ni' \quad . \quad . \quad . \quad . \quad (77)$$

was capable of almost innumerable applications. It is true that, owing to the want of accurate information on the behaviour of polyatomic gases as regards their specific heat at low temperatures, these applications largely took the form of approximate estimates. Nevertheless, the fact that the results of these estimates agreed well with the truth in every case was certainly very much in favour of the new Heat Theorem, even before the numerous exact proofs described in Chapter IX were available. The approximation formula to be given below is extremely simple in its application, and can also serve as a control both on experiment, at least in so far as concerns very large deviations, and also on theoretical speculations. We shall hence be justified if we deal with it in some detail.

It should be mentioned that some erroneous tendencies in recent chemical literature render the following warning necessary. The approximation formula must not be identified with my Heat Theorem, although it was found in connection therewith. Discrepancies between the approximation formula and the facts do not of course entail any contradiction of the strict validity of the Heat Theorem. This would only come in question if a discrepancy were found between an exact application of it and the facts; such has never been the case hitherto, and is now hardly to be expected in view of the present vastly increased state of our knowledge.

2. Derivation of a Vapour-Pressure Formula.—The following is the method which we select in order to find a useful approximation formula. We attempt first to obtain an equation correct at least in its outlines; for, as we do not know the specific heats in both the liquid and gaseous states at low temperatures, we cannot obtain a vapour-pressure equation which is exact, i.e. valid to near the absolute zero. By applying such equations to the components of a condensed system, as shown in the last chapter, we then obtain a relation which holds for gaseous chemical systems.

We have previously observed (p. 124) that at sufficiently low temperatures the specific heats of all solids and also of liquids must be less than the specific heat of the vapour. According to the formula

$$\frac{d\lambda}{dT} = C_p - c \qquad . \qquad . \qquad . \quad (79)$$

(C_p is the molecular heat of the vapour at constant pressure and c the molecular heat of the liquid), it follows that the heat of evaporation must first rise at very low temperatures, and only decrease again at higher temperatures, becoming zero at the critical point.

The fact that the heat of evaporation must always have a maximum, which has now been established beyond a doubt by many investigations, was quite unknown when I propounded my Heat Theorem, and was first deduced by me theoretically by its means. A vapour-pressure formula which is to correspond as completely as possible with the vapour-pressure curve instead of being valid, like the numerous interpolation formulæ hitherto proposed, over a limited range of temperature only, must naturally take this circumstance into account

We take as a formula which correctly reproduces the course of the heat of vaporization, at least in its main features,

$$\lambda = (\lambda_0 + \alpha T - \epsilon T^2)\left(1 - \frac{p}{\pi_0}\right) \qquad . \quad . \quad (80)$$

The factor $\left(1 - \frac{p}{\pi_0}\right)$, where π_0 = critical pressure, will not differ appreciably from 1 until the vapour pressure is high;

it has the effect, however, that λ vanishes at the critical point. The coefficients a and ϵ are positive; the two terms in which they occur condition an initial rise and a later decrease in the heat of evaporation.

We are thus sure that the above equation reproduces the variation of the heat of evaporation not, of course, accurately, but, at any rate, in its general outlines; in contrast with all formulæ hitherto proposed, it does so from the lowest temperatures up to the critical point.

By substituting for λ in the well-known equation

$$\lambda = T\frac{dp}{dT}(v - v'), \qquad . \qquad . \qquad . \qquad \textbf{(81)}$$

and integrating, we obtain the desired vapour-pressure formula.

The integration is only rendered possible by the following relation. I found, on calculating from Young's measurements on fluor-benzene, an equation,*

$$p(v - v') = RT\left(1 - \frac{p}{\pi_0}\right). \qquad . \qquad . \qquad \textbf{(82)}$$

valid up to fairly high pressures. Since the theorem of corresponding states holds fairly well for volume relations, in particular for the deviations from the gaseous state, it is probable that the foregoing equation may serve generally as a good approximation. We shall doubtless come nearer to the truth by using it than if we neglect v' compared with v in equation **(81)**, as is usually done, and apply the gas laws to the vapour. The latter procedure, of course, renders it impossible to obtain a formula valid up to pressures which are at all high.

Integration of **(81)**, making use of **(80)** and **(82)**, gives

* Cf. also the recent paper by Schimank (104). [The equation has been shown by Herz, " Zeits. f. Elektrochem.," **30**, 604, 1924, to hold for a large number of substances up to 0·8-0·9 of the critical temperature—Tr.]

$$\lambda = RT^2 \frac{d \log_e p}{dT}\left(1 - \frac{p}{\pi_0}\right) \quad . \quad . \quad (83)$$

$$\log_e p = -\frac{\lambda_0}{RT} + \frac{a}{R} \log_e T - \frac{\epsilon}{R}T + i \quad . \quad (84)$$

If we assume that at very low temperatures the molecular heat of all gases converges towards $\frac{5}{2}$ R, and that the molecular heat of the liquid becomes negligibly small, a will take (cf. also p. 124) the generally applicable value

$$a = 2 \cdot 5R = 4 \cdot 96.$$

At the time when I was trying to find the approximation formula, both of these assumptions, which may at the present time be considered warranted, appeared very risky. I did assume (1), however, that a had a value independent of the nature of the particular substance, and I determined [*] the most probable value as

$$a = 1 \cdot 75R = 3 \cdot 5,$$

as the result of many tedious calculations.

Even to-day it does not seem to me at all certain that it is desirable to introduce the theoretical value of a into equation **(84)**. The specific heat of the liquid even at extremely low temperatures is an appreciable amount; the theoretical value must therefore decrease very rapidly, and it is consequently quite conceivable that a somewhat smaller value (3·5 instead of 4·96) should better reproduce the actual facts.

If we now introduce common logarithms into equation **(84)** we have finally

$$\log_{10} p = -\frac{\lambda_0}{4 \cdot 571 T} + 1 \cdot 75 \log_{10} T - \frac{\epsilon}{4 \cdot 571} T + C \quad (85)$$

$$\lambda = (\lambda_0 + 3 \cdot 5T - \epsilon T^2)\left(1 - \frac{p}{\pi_0}\right) \quad . \quad . \quad (86)$$

---

[*] Cf. also the calculations of E. C. Bingham (5a).

APPROXIMATION FORMULA

In the case of oxygen, for example, H. v. Siemens (91) found the equation

$$\log_{10} p = -\frac{399}{T} + 1{\cdot}75 \log_{10} T - 0{\cdot}01292 T + 5{\cdot}0527$$

where p is expressed in mm. of Hg :—

T	57·37	59·21	59·98	63·08	67·20	72·63	76·02	80·31	85·45	90·18
p (obs.)	2·68	4·40	5·49	11·52	28·07	75·7	129·5	239·5	457·6	766·8
p (calc.)	2·72	4·48	5·46	11·53	27·83	75·4	129·8	239·8	456·7	766·5

The concordance between calculation and experiment is excellent; the heat of evaporation is hence calculated as 1624 and 1784 for the temperatures 90·1° and 68·2° respectively, whereas Alt found 1629 and 1777.

In numerous other cases the above vapour-pressure formula has shown itself far superior to any of the corresponding expressions obtained before, in so far as concerns the calculation over large intervals of temperature with the fewest possible constants.*

Equations **(85)** and **(86)** contain three specific constants; but it is to be noted that C has the same value for a given substance no matter whether the vapour pressure concerned is that of the liquid or of the crystallized state; and we shall learn later to know certain rules for C, which allow us to derive at least an approximate value for it from other data.

3. Approximation Formula for Chemical Equilibria.—We now return to the subject-matter of page 125; instead of operating with the vapour-pressure formula,

$$\log_e \xi = -\frac{\lambda_0}{RT} + \left(\frac{a_0}{R} - 1\right)\log_e T + \frac{\beta_0}{R}T + \frac{\gamma_0}{2R}T^2 + \ldots + i \tag{67}$$

we use the approximation formula **(85)**,

$$\log_{10} p = -\frac{\lambda_0}{4{\cdot}571 T} + 1{\cdot}75 \log_{10} T - \frac{\epsilon}{4{\cdot}571}T + C \tag{85}$$

where we have put

$$C = \frac{i + \log_e R}{2{\cdot}3023}.$$

---

* For further examples, see Brill (4), Naumann (11), E. Falck (17), Nernst (25), Baker (30), Mündel (92).

If, further, as is usually preferable for gaseous systems, we work with partial pressures instead of with concentrations, and put, accordingly,

$$K' = \frac{p_1^{\nu_1} p_2^{\nu_2} \cdots}{p_1'^{\nu_1'} p_2'^{\nu_2'} \cdots} = K \cdot (RT)^{\nu_1 + \nu_2 + \cdots - \nu_1' - \nu_2' - \cdots} \quad (86)$$

then it may easily be shown that

$$\log_{10} K' = -\frac{Q_0}{4 \cdot 571 T} + \Sigma \nu\, 1 \cdot 75 \log_{10} T + \frac{\beta}{4 \cdot 571} T + \Sigma \nu C \quad (87)$$

The heat evolution, Q, at constant pressure is consequently given by the expression

$$Q = Q_0 + \Sigma \nu\, 3 \cdot 5\, T + \beta T^2 \quad . \quad . \quad (88)$$

In the foregoing formulæ we have restricted ourselves to the first two terms of the developed series, as was also done in deriving the approximate vapour-pressure formula.

4. "Conventional Chemical Constants."—The problem now arose to determine the values of C for as many gases as possible; we may obviously regard C as a property of the particular gas, because it must be independent of the nature of the condensate, and therefore depends only on the nature of the gas.

The most direct method is, of course, to derive it from vapour-pressure measurements, but this method is difficult, as was shown particularly by Mündel (92), in that vapour-pressure curves, especially at low pressures, may be represented fairly well by making fairly large simultaneous variations of the three values λ, ϵ, and C. At high temperatures, i.e. near the critical point, we have to a certain approximation, as was shown by van der Waals in 1881,

$$\log_{10} \frac{\pi_0}{p} = a \left(\frac{\theta_0}{T} - 1\right) \quad . \quad . \quad (89)$$

where π_0 and θ_0 are the critical pressure and temperature. There is an apparent parallelism between the constant a contained in this equation—which is, of course, only a con-

APPROXIMATION FORMULA

stant in the neighbourhood of the critical point—and the quantity C, of which we are in search; both constants, when arranged in order of magnitude, are in the same sequence.

In my calculations at the time I came across a then unknown property of what is known as Trouton's constant $\frac{\lambda}{T_0}$ (heat of evaporation at the boiling-point divided by the absolute temperature of the boiling-point). It was known that for the so-called normal, i.e. non-associated, liquids the constant had a value about 22. If liquids of very low boiling-point such as nitrogen or hydrogen are so compared a regular decrease is found; nitrogen gives 17·6, hydrogen 12·2, and helium, according to the recent results of Kamerlingh-Onnes, only 5·1 instead of 22.*

There was again a clear parallelism between the values of C and $\frac{\lambda}{T_0}$, and we may put

$$C = 1\cdot 1a = 0\cdot 14\frac{\lambda}{T_0} \text{ (approximately!)} \quad . \quad (90)$$

It is to be noted that such a relation can only be purely accidental, in so far as concerns the direct proportionality: a is independent of the units employed, but C and $\frac{\lambda}{T_0}$ depend on them, and depend in different ways. If we alter the unit of pressure, another term is added to C. No deep meaning can therefore be ascribed except to the relation, which is independent of the units selected, that the sequence of the three quantities, C, a, and $\frac{\lambda}{T_0}$ is the same, or, at any rate, nearly the same.

Cederberg † has recently drawn attention to a third relation in a very interesting paper, about which we shall

---

* Cf. " Theor. Chem.," p. 319.

† J. W. Cederberg, " Thermodyn. Berechnung chem. Affinitäten, Dissertation," Upsala, 1916.

have more to say later. According to this, the sequence of C and $\log \pi_0$ is also essentially * the same, and we may put

$$C = 1.7 \log_{10} \pi_0 \quad . \quad . \quad . \quad (91)$$

In the following table are collected the values of C for the most important substances with which we are concerned. Since these values, in conjunction with the thermal properties, characterize the chemical behaviour of the respective molecular species, I have called them " chemical constants "; but as the values here given have been calculated on certain assumptions which are only approximately true, we shall term them " conventional chemical constants," in contrast with the " true chemical constants " calculated in Chapter XIII.

TABLE VIII
Conventional Chemical Constants

He	0·6	I_2	3·9	CO_2	3·2
H_2	1·6	HCl	3·0	CS_2	3·1
CH_4	2·5	HI	3·4	NH_3	3·3
N_2	2·6	NO	3·5	H_2O	3·6
O_2	2·8	N_2O	3·3	CCl_4	3·1
CO	3·5	H_2S	3·0	$CHCl_3$	3·2
Cl_2	3·1	SO_2	3·3	C_6H_6	3·0

We notice that the majority of substances have a value of C lying in the neighbourhood of 3·1; the low-boiling substances, in particular hydrogen and helium, show smaller values, while associated substances have higher values.

For gases not included in this table a value may readily be obtained from one of the above rules, or it may be estimated, if necessary, from the molecular weight.

It should be observed that, as required by the nature of the case, it is never C alone, but always the combination $1.75 \log_{10} T + C$ which occurs in all thermodynamical calculations. If we replace the factor 1·75 by one rather smaller or greater, which will only slightly influence the utility of the vapour-pressure formula, we arrive at somewhat

---

* For NO the relation does not hold; cf. p. 142.

APPROXIMATION FORMULA

different values, naturally, for the chemical constants, which may also give good results. This is really only another expression of the fact that there may be interpolation formulæ of different constructions which can nevertheless be used quite well. We should not forget either that, starting from condensed systems, the new Heat Theorem can be tested by means of any vapour-pressure formula whatsoever, provided that it agrees with the observations.

In view, particularly, of our inadequate knowledge of the behaviour of the specific heats of polyatomic gases, it will only rarely be possible to calculate the coefficients β or higher terms; consequently the following still more simplified approximation formula may be employed temporarily:

$$\log_{10} K' = - \frac{Q'}{4 \cdot 571 T} + \Sigma \nu \, 1 \cdot 75 \log_{10} T + \Sigma \nu C. \quad (92)$$

Here Q' is the heat evolution at constant pressure for the ordinary temperature, as obtained directly from the thermo-chemical tables.

In dealing with a test of the Heat Theorem here developed one must always, naturally, go back to the fundamental equations; but if one wishes to obtain quickly, with the aid of the thermo-chemical tables, an idea of the *approximate* position of an equilibrium, the above simple equation does, in fact, appear to be very suitable, as calculation has shown in numerous cases.

A few examples of this sort will be mentioned in the following section; but before doing so, it will be useful to make one or two remarks as a guide to the difference which there must be between the capabilities of the more general and of the simplified approximation formulæ. For this purpose we shall apply the two formulæ to the transformation of monoclinic to rhombic sulphur, with which we dealt on page 106.

The affinity of the transformation per mol may be written in this case,

$$A = RT \log_e \frac{\pi_1}{\pi_2},$$

where π_1 is the sublimation pressure of the monosymmetric and π_2 that of the rhombic modification.

If, according to the general approximation formula, we put

$$\log_e \pi_1 = -\frac{\lambda_0'}{RT} + 1{\cdot}75 \log_e T + \frac{\epsilon'}{R}T + C,$$

$$\log_e \pi_2 = -\frac{\lambda_0''}{RT} + 1{\cdot}75 \log_e T + \frac{\epsilon''}{R}T + C,$$

we obtain for A

$$A = \lambda_0'' - \lambda_0' - (\epsilon'' - \epsilon')T^2,$$

and thus for U,

$$U = \lambda_0'' - \lambda_0' + (\epsilon'' - \epsilon')T^2.$$

Now this is the approximation already discussed on page 106, which reproduces the actual facts, not by any means exactly, it is true, but still with some degree of approximation, and it has certainly given us something which goes far to explain the general thermodynamical behaviour of an allotropic change.

Following the simplified approximation formula, we put

$$\log_e \pi_1 = -\frac{\lambda_0'}{RT} + 1{\cdot}75 \log_e T + C,$$

$$\log_e \pi_2 = -\frac{\lambda_0''}{RT} + 1{\cdot}75 \log_e T + C.$$

This gives simply

$$U = A = \lambda_0'' - \lambda_0',$$

a result which is in fact quite unsatisfactory and insufficient. It may be shown that, in general, equilibria like the melting-point or transition-point, are not within the scope of the simplified formula; applied to condensed systems it merely becomes comparable with Berthelot's rule.

Nevertheless it is, as we shall shortly see, much to be preferred to the latter in that it allows an excellent expression of the nature of equilibria in which a gaseous phase takes part.

There is no doubt that it would be a great advance if it

APPROXIMATION FORMULA

were not necessary to neglect the ϵ coefficients; we could then conceivably deal with the values of ϵ just as we did with the values of C, and give a table and, above all, general rules for a simple determination of them. It is probable that this may be done, even though the problem here is more complicated, because such a determination would have to extend not only to gases, but also to the condensates from them; in the case of water, for example, the value of ϵ would have to be determined for the gas, for water, for ice, for water of crystallization (and for each hydrate separately). Since it is only a species of correction term which is involved, it is possible that the problem may be simplified in practice, and it would perhaps be worth while to undertake work with this object in view. On the other hand, the preparation of tables for U and A, such as are cited on page 97, goes much further, for it entails a solution of the problem which is exhaustive and neglects nothing, though it does not, in the first place, take account of gases.

5. Applications of the Simplified Approximation Formula.—Let us consider first the case when $\Sigma \nu = 0$, i.e. when there are just as many gaseous molecules disappearing as being formed in the reaction; then

$$\Sigma \nu \cdot 1.75 \log T = 0,$$

and it is further to be noted that the expression $\Sigma \nu \cdot C$ is not very important, since the values of C are nearly equal. If the heat evolution in this case is also equal to nil, it follows from equation **(92)** that

$$\log K' = 0.$$

If, as in the transition from one optical isomer into its antipode, there are only two molecular species concerned,

$$\log \frac{p}{p'} = 0, \; p = p';$$

in other words, the two antipodes when in equilibrium must have equal partial pressures. This result was, of course, deduced by van't Hoff long ago from kinetic considerations, but it had not hitherto been possible to demonstrate it from thermodynamics. In view of the equality of vapour pressure of the two modifications, it may be presumed from equations **(73)** and **(77)** that the above result, and in this case our approximation formula also, is rigidly true.

Conditions are similar in the reaction

$$H_2 + I_2 = 2HI \ldots + 2760 \text{ cals.}$$

Here we have to expect, from the smallness of the heat change, that an equilibrium should be obtained even at low temperatures, and that, though it must be displaced in the direction of the formation of hydriodic acid, yet all the molecular species take part in the equilibrium with appreciable partial pressures. Experience confirms this fully. In addition to the examples mentioned, we may recall the formation of an ester, the classical case of chemical equilibrium; since the formation of gaseous ethyl acetate and water from alcohol and acetic acid vapours takes place without any considerable heat change, an equilibrium must be produced just as with hydriodic acid, in which all the components participate in considerable concentrations; since the vapour pressures of the four substances are not very different, this equilibrium must also be found for the liquid mixture.

It is well known how important a rôle has been played in the theory of chemical affinity by investigations on the ester equilibrium (Berthelot and Péan St. Gilles) and on the dissociation of hydriodic acid (Lemoine and Bodenstein). As we now know, this is no accident, for in both cases reactions are concerned which must, on thermodynamic grounds, show marked evidence of the equilibrium state at quite low temperatures.

In the reactions in the homogeneous gaseous systems,

$$H_2 + Br_2 = 2HBr \ldots + 24{,}200 \text{ cals.}$$
$$H_2 + Cl_2 = 2HCl \ldots + 44{,}000 \text{ cals.},$$

APPROXIMATION FORMULA

we may expect, from the amount of the heat evolution, that dissociation should become appreciable only at temperatures much higher than for hydriodic acid. It may be seen that the results coincide here qualitatively with Berthelot's Principle; from a quantitative aspect the superiority of our approximation formula is evident in that the extent of the dissociation may also be roughly estimated by its means, as I have shown in detail (25) for the equilibrium of the three halogen hydrides.

In this connection a comparison of the reactions,

$$H_2 + Cl_2 = 2HCl \ldots + 44{,}000 \text{ cals.,}$$
$$2NO = N_2 + O_2 \ldots + 43{,}200 \text{ cals.,}$$
$$CO + H_2O = H_2 + CO_2 \ldots + 10{,}200 \text{ cals.,}$$

is particularly instructive.

Neglecting the specific heats, i.e. the variability of Q which they cause, the Second Law gives for the above cases

$$\log_{10} K = -\frac{Q}{4 \cdot 57 T} + I,$$

and our approximation formula requires
$$I = \Sigma \nu C.$$

If, in order to calculate C without any arbitrariness, we introduce in each case the relation

$$C \backsim 0 \cdot 14 \frac{\lambda}{T_0}$$

we obtain

$$\log_{10} K = -\frac{Q}{4 \cdot 57 T} + 0 \cdot 14 \Sigma \nu \frac{\lambda}{T_0}.$$

In the following table are given the values of what are known as "Trouton's constants," so far as they are required for the above three reactions: they are calculated, both directly from the observed heats of evaporation and boiling-points and also from a very interesting formula given by Cederberg (*loc. cit.*, p. 55),

$$\frac{\lambda}{T_0} = \frac{4 \cdot 57 \log_{10} \pi_0}{1 - \frac{T_0}{\theta_0}} \left(1 - \frac{1}{\pi_0}\right) \quad . \quad . \quad (93)$$

As may be readily seen, this formula bears a certain relationship to that **(90)** previously found by me (p. 135); for if we put $p = 1$ atm. in equation **(89)**, **(93)** becomes

$$\frac{\lambda}{T_0} = 4.57\, a\left(1 - \frac{1}{\pi_0}\right)\frac{\theta_0}{T}; \qquad . \quad . \quad \textbf{(94)}$$

and if we observe that $\dfrac{T_0}{\theta_0}$ for most substances is about 0.6, then **(90)** and **(94)** become practically identical.

TABLE IX

Trouton's Coefficients

	T_0	λ	θ_0	π_0	$\frac{\lambda}{T_0}$ found.	$\frac{\lambda}{T_0}$ calc.	Mean.
H_2	20.4	248	32.0	11.0	12.1	11.9	12.0
O_2	90.1	1642	154.3	49.7	18.2	18.3	18.2
N_2	77.2	1362	126.0	33.5	17.6	17.5	17.5
CO	83.0	1414	134.4	34.6	17.0	17.9	17.5
NO	122.9	3000	180.2	64.6	24.4	25.5	25.0
CO_2	187	4420	304.2	73.0	23.6	21.8	22.7
HCl	190.2	3600	324.5	81.6	19.0	20.7	19.8
Cl_2	239.5	4500	419.1	93.5	18.8	20.7	19.7
H_2O	373.1	9650	647.1	217.5	25.9	25.1	25.5

Cederberg's formula holds very well for the observations here used, as may be seen from a comparison of the two columns before the last; the following remarks may be made about certain cases in the above table. The molecular heat of evaporation for NO has been calculated from the vapour-pressure measurements by Adwentowski,[*] but the different measurements give values so discordant that the figure so obtained can only be approximate; in any case, it is certain, if we take into consideration also the values of Trouton's coefficient obtained according to Cederberg, that NO has an exceptionally high value of $\dfrac{\lambda}{T_0}$, or in other words, is a strongly associated substance.[†] For carbon dioxide, which of course freezes before the vapour pressure has fallen

---

[*] " Chem. Zentralblatt," 1910, I, p. 1106.
[†] See also Supplement, p. 265.

to atmospheric pressure, λ and T_0 were calculated for the liquid state by Frl. Falk (17) from the equations of state; the values may be considered quite reliable. Here again, in agreement with Cederberg's formula, association, though only moderate, is to be presumed, since the normal value of Trouton's coefficient for the boiling-point $T_0 = 187$ amounts to only 20·3.

The equilibria of the above three reactions in the neighbourhood of $T = 1000°$ may be represented by the formulæ—

$$\log_{10} K = -\frac{44000}{4\cdot 57\,T} - 0\cdot 70 \qquad (-\,1\cdot 11)$$

$$\log_{10} K = -\frac{43200}{4\cdot 57\,T} + 1\cdot 08 \qquad (+\,1\cdot 60)$$

$$\log_{10} K = -\frac{10200}{4\cdot 57\,T} + 2\cdot 05 \qquad (+\,1\cdot 16).$$

Calculation from the formula $I = \Sigma \nu C$ gives for the integration constant I the results appended in the brackets.

There is, of course, nothing like numerical agreement and, in view of the character of the approximation formula, it was not to be expected. But if equilibria of this sort are required to be calculated from thermodynamical data, it is naturally very valuable to have a simple method of obtaining an idea, at any rate, of the sign and of the approximate magnitude of the hitherto quite unknown constant of integration I; this, as the above examples show, the approximation formula does provide. For the formation of hydrobromic and hydriodic acids the conditions are quite analogous to those for the formation of hydrochloric acid; of this it is very easy to convince one's self (cf. 25).

The following remarks may illustrate the nature of our approximation formula. In the first of the three reactions given on page 141, the unusually low value of the chemical constant of hydrogen on the left side of the equation indicates that the constant of integration I is negative. In the second reaction there are on the left-hand side two molecules of strongly associated NO, and consequently a high positive value results for I. The same thing occurs in the third reaction, for on the left there is the strongly associated water, and on the right the hydrogen with its small chemical constant.

The relations between chemical equilibrium and Trouton's coefficient are therefore unmistakable, and may be employed with advantage. We must not, however, lose sight of the fact that the equations with which we are dealing are only approximately true. The exact treatment of the problem, which depends on the use of the specific heats of gases down to very low temperatures, will be learned in Chapters XIII and XIV, where it will also be seen that these specific heats are mostly inaccessible to experiment at the present time. This is the reason why I have thought it necessary to devote so much space to the treatment of the approximation formula.

Compounds, e.g. NO, which are formed with a large absorption of heat without a change in the number of molecules, are not stable until the temperature is high; we have here in general

$$\log_{10} K' = -\frac{Q'}{4 \cdot 57\,T} + \Sigma \nu C;$$

from this equation it may be seen that so long as Q' is large compared with $4 \cdot 57T$, only a small percentage of the compound is capable of existence in equilibrium, for $\Sigma \nu C$, even in extreme cases, is only a little greater than 1, and as a rule is actually less than 1, owing to the approximate equality of the values of C.

A further example of the same sort is the formation of acetylene from hydrogen and solid carbon according to the equation,

$$H_2 + 2C = C_2H_2,$$

which takes place, like the formation of nitric oxide, with the absorption of much heat; corresponding with this, acetylene does not begin to be formed from its components

APPROXIMATION FORMULA

until the temperature is high, and even then only in minimal quantities (cf. v. Wartenberg, 6).

Carbon dioxide is formed with a large evolution of heat from solid carbon and oxygen :

$$(a) \quad C + O_2 = CO_2 + 94,260 \text{ cals.}$$

(referred to graphite) ; the maximum work, when the CO_2 formed has the same partial pressure as the oxygen reacting, amounts here to

$$A = 4\cdot 571 T \log_{10} K'.$$

It would be of the greatest interest to calculate the affinity in this case, for the reason that one would then know what is the maximum work which can be developed by this reaction, which operates the majority of our machinery. I indicated a method of doing this in the first edition of my text-book (1893) : for if we know the equilibria both of the reaction

$$(b) \quad C + CO_2 = 2CO,$$

and of the reaction

$$(c) \quad 2CO_2 = 2CO + O_2,$$

we may determine from (b) and (c) the equilibrium of the reaction (a), and hence also its affinity.

On carrying out this calculation * it was found that A and Q' are here only slightly different. The same result is given immediately by applying our approximation formula direct to reaction (a),

$$A = Q' + 4\cdot 57 T(2\cdot 8 - 3\cdot 2),$$

the second summand being negligible compared with Q'.†

Let us now consider the case where an alteration takes

---

* Nernst, " Theor. Chem." 1 Aufl., p. 545 (1893) ; English translation of 8th-10th ed., p. 800.

† Analogous calculations for other fuels are given in Pollitzer's monograph.

place in the number of molecules of the gaseous components of the reaction; the classical examples of this are the phenomena of dissociation. In this case the sum of the last two terms of equation **(92)** is a positive quantity, and is by no means small, especially at high temperatures; this explains why, although the reaction

$$2NO_2 = N_2O_4,$$

for example, takes place with the evolution of the considerable heat of 12,500 cals., an equilibrium marked with dissociation occurs in the neighbourhood of T = 300, and why, at slightly higher temperatures, N_2O_4 decomposes almost quantitatively.

In general, the values of C will not be known; e.g. in the above case we cannot apply the methods described on page 135 to the molecules NO and N_2O_4. We obtain, however, a fairly rough idea if we equate the values of C to 3 as an approximation. Representing by x the extent of dissociation when the total pressure is P atm., we have

$$\log_{10} K' = \log_{10} \frac{x^2}{1-x^2} P = -\frac{Q'}{4 \cdot 57 T} + 1 \cdot 75 \log_{10} T + 3.$$

Here, in contrast with the examples of pages 139 and 143, the value which has to be added to the first term on the right-hand side is, as we have already indicated, not inconsiderable, and is, moreover, of the opposite sign; thus, according to our formula, we must be encountering a totally different behaviour, namely, a complete breakdown of Berthelot's Principle; this has been fully confirmed quantitatively. When

$$x = 0 \cdot 5, \ K = \frac{P}{3},$$

so that we obtain for the temperature at which the dissociation has proceeded halfway the expression

$$T_1 = \frac{Q'}{4 \cdot 57 \, (\log_{10} 3 + 1 \cdot 75 \log_{10} T_1 + 3 - \log_{10} P)}.$$

APPROXIMATION FORMULA

Given Q' and P, T_1 may be readily evaluated by trial and error, since the term $1.75 \log_{10} T_1$ on the right side varies only slowly with the temperature.

The following table has been calculated by Brill (7) and Pollitzer (Monograph, p. 102).

TABLE X
Dissociation of Gases

Reaction.	Q'.	P.	T_1 (obs.)	T_1 (calc.)
$2NO_2 = N_2O_4$	12,450	0.65	323	340
$2HCOOH = (HCOOH)_2$	14,780	1	410	410
$2CH_3COOH = (CH_3COOH)_2$	16,600	1	425	450
$PCl_3 + Cl_2 = PCl_5$	18,500	1	480	500
$HBr + C_5H_{10} = C_5H_{10} \cdot HBr$	19,400	1	483	525
,, ,, = ,, ,,	19,400	0.1	462	470
$H_2O + SO_3 = H_2SO_4$	21,850	1	623	599

In the following table are collected the values of T_1, which correspond with the heats of dissociation given above them :—

Q'	10,000	20,000	50,000	100,000	200,000
T_1	290	525	1,220	2,350	4,500

According to the result first obtained by me and confirmed later by v. Wartenberg, the dissociation

$$S_2 = 2S$$

at 2320° abs., and atmospheric pressure was found by method 4 (p. 16) to be *ca.* 50 per cent., whence the heat of dissociation is calculated to be *ca.* 90,000 cals., whereas Budde (67) and Bjerrum (73) were led to values of 80,000 to 120,000 by the explosion method. The dissociation

$$Cl_2 = 2Cl,$$

measured by Pier by the same method, showed a heat of dissociation of 113,000 cals., and chlorine at 2610° abs. was found to be half split up into atoms. Russel,* who recently repeated this experiment with more refined methods, found the value 106,000 cals., which is close to the above, and agrees well with that calculated according to the approximation formula. According to our results, obtained by the explosion method, oxygen and nitrogen begin to dissociate only at temperatures very much higher than those for the gases just mentioned; corresponding with this, values of the heat of dissociation lying much above 100,000 cals. are to be assumed. In the case of hydrogen, signs of the commencement of dissociation could be recognized, by the explosion method, above 2000° (Lummerzheim †); taking into account the fact that H_2 and *a fortiori* H have values of C considerably less than 3, the heat of dissociation may be estimated to amount to some 100,000 (for further details, cf. Chapter XIII).

In the dissociation of many solids, e.g. of calcium carbonate,

$$CaO + CO_2 = CaCO_3,$$

we again meet with the case where there is a molecule more of gas on the left side of the equation than on the right; in the light of our approximation formula this reaction must exhibit a certain analogy with the phenomena of dissociation discussed above. In fact, if we put C = 3·2 for CO_2, we obtain for the dissociation of calcium carbonate or similar processes,

$$\log_{10} p = - \frac{Q'}{4 \cdot 571 T} + 1 \cdot 75 \log_{10} T + 3 \cdot 2,$$

or, if we wish to calculate the temperature T_1 at which the dissociation pressure becomes an atmosphere,

$$\frac{Q'}{T_1} = 4 \cdot 57 (1 \cdot 75 \log_{10} T_1 + 3 \cdot 2) \quad . \quad . \quad (94a)$$

* The results of Russel and of Lummerzheim cannot yet be published.
† Ibid.

APPROXIMATION FORMULA

For values of T which lie in the neighbourhood of 250° to 350°, we have approximately, since log T varies only slowly with the temperature,

$$4 \cdot 57(1 \cdot 75 \log_{10} T_1 + 3 \cdot 2) = ca.\ 34,$$

and thus

$$T_1 = \frac{Q'}{34}.$$

As a matter of fact, this rule was found empirically by le Chatelier and Matignon, who proposed for the sublimation and dissociation of solids the law

$$\frac{Q'}{T_1} = ca.\ 33 \quad . \quad . \quad . \quad \textbf{(94}b\textbf{)}$$

Our approximation formula not only confirms this empirical discovery within certain limits, but renders it more precise and extends it in many ways :—

1. The empirical value 33 has now received a simple theoretical significance, and there is a definite temperature-effect to be expected. The following table by Brill (7) shows the superiority of our new formulation over the rule of le Chatelier-Matignon :—

TABLE XI

Dissociation of Solids *

Substance.	Q'.	T_1 calc. from (94b).	T_1 calc. from (94a).	T_1 obs.
Ag_2CO_3	20,060	627	548	498
$PbCO_3$	22,580	706	610	575
$MnCO_3$	23,500	741	632	$ca.$ 600
$CaCO_3$	42,520	1329	1091	1098
$SrCO_3$	55,770	1743	1403	1428

* In my Silliman lecture (1907) the corresponding table on p. 109 has been put in quite the wrong order by an oversight : the headings were introduced from another page of the proof, and may be misleading.

2. At extremely high and at extremely low temperatures a complete breakdown of the purely empirical rule is to be expected according to our approximation formula. Two cases of this sort, which are particularly telling, have been calculated out by Winternitz (98*a*), namely, the sublimation pressure of solid tungsten and that of solid hydrogen.

In the former case

$$\frac{Q'}{T_1} = \frac{209000}{5100} = 41,$$

while the approximation formula gives

$$(1\cdot75 \log_{10} T_1 + 3)4\cdot57 = 42;$$

in the second case

$$\frac{Q'}{T_1} = \frac{279}{19} = 15,$$

the approximation formula requiring

$$(1\cdot75 \log_{10} T_1 + 1\cdot6)4\cdot57 = 18.$$

Both values deviate very considerably from the "normal value" of 33, and, as was to be foreseen, in opposite senses. Our approximation formula, however, gives an agreement which is at least moderately good in both cases. The observed value is only of medium accuracy in either case.

3. Lastly, the regularity with which we are dealing is now capable of considerable extension; for we may lay down rules of the same sort for any reaction in which the number of gaseous molecules is different on the two sides of the equation to the reaction. The formula on page 146 gives, if we put $P = 1$, a corresponding rule for the break-up of a gaseous molecule into two others, and Table X corroborates it. We shall not here trouble to lay down further rules of this kind, because in any given case the direct application of our approximation formula will be preferable owing to its greater accuracy.

As a last example, let us consider the case where several gases result from the dissociation of a solid.

APPROXIMATION FORMULA

For the dissociation of the nitrates of bivalent metals, taking place, for example, according to the equation

$$Pb(NO_3)_2 = PbO + 2NO_2 + \tfrac{1}{2}O_2,$$

formula **(92)** takes the form (with $P = 1$)

$$\frac{Q'}{T_1} = 2\cdot 5 \times 4\cdot 57(1\cdot 75 \log_{10} T + 3);$$

i.e. equation **(94b)**, page 149, does not hold even approximately; this agrees with experience and explains very simply the fact, which must have puzzled Thomsen (Pollitzer, p. 118) that the nitrates, in spite of their greater heat of formation, dissociate at lower temperatures than the carbonates.

A similar case has recently been dealt with by Bodenstein;* the reduction of zinc oxide by carbon takes place at approximately the same temperature as that of potassium carbonate by carbon, although much more heat is absorbed in the latter case than in the former; according to our formulæ, this is simply explained in that many more gas molecules are set free in the second case than in the first.

To sum up, the approximation formula **(92)** gives us a good general picture of what takes place in heterogeneous equilibria, where gases participate in the reaction; the deviations from Berthelot's Principle become the greater the more the influence of the second term in the equation

$$\log_{10} K' = - \frac{Q'}{4\cdot 57 T} + \Sigma\nu(1\cdot 75 \log_{10} T + C)$$

makes itself felt. In a reaction such as

$$C + O_2 = CO_2$$

this influence is almost negligible, and thus the heat evolution is here a fairly accurate measure of the affinity (cf. p. 145). In a reaction such as

$$CaCO_3 = CaO + CO_2,$$

* "Zeitschr. f. Electrochemie," 1917, p. 103.

where there is one more gaseous molecule on the right side of the equation than on the left, the reaction may proceed endothermally at high temperatures with great chemical energy; much more is this the case for reactions such as

$$NH_4Cl = NH_3 + HCl,$$

in which the number of gaseous molecules split off from the solid condensate is still higher.

6. Cederberg's Approximation Formula.—In a very interesting and noteworthy paper, J. W. Cederberg [*] puts forward an approximation formula slightly different from that which I have derived and used above. In place of my approximate vapour-pressure formula,

$$\log_{10} p = -\frac{\lambda_0}{4\cdot 57 T} + 1\cdot 75 \log_{10} T - \frac{\epsilon}{4\cdot 57}T + C. \quad (95)$$

he puts

$$\log_{10} p = -\frac{\lambda_0}{4\cdot 57 T} + 2\cdot 5 \log_{10} T - \frac{\epsilon}{4\cdot 57}T^{7/4} + \log_{10} \pi_0. \quad (96)$$

Numerous examples show that the latter formula also reproduces the measurements extremely well, in some cases over an extraordinarily wide range of temperature.

The two formulæ give respectively for the heats of evaporation—

$$\lambda = (\lambda_0 + 3\cdot 5 T - \epsilon T^2)\left(1 - \frac{p}{\pi_0}\right) \quad . \quad (97)$$

$$\lambda = (\lambda_0 + 5\cdot 0 T - \tfrac{3}{4}\epsilon T^{7/4})\left(1 - \frac{p}{\pi_0}\right) \quad . \quad (98)$$

As concerns the replacement of the coefficient 1·75, empirically deduced by me as the best approximation, by the theoretical value 2·5, I must point out that there is not necessarily, as Cederberg states on page 22, any practical advantage in this (cf. p. 132). The approximation formula has not hitherto been used at temperatures so low that the

[*] Dissertation, Upsala, 1916, "Thermodyn. Berechnung chem. Affinitäten."

APPROXIMATION FORMULA

use of the theoretical factor would be obligatory, and it might thus be quite possible that my empirically-found coefficient is preferable in the region for which the approximation formula is, in the nature of things, intended. Moreover, it has been overlooked that the theoretical basis of the coefficient 2·5 was first given by me (cf. p. 124) after I had proposed the two general rules :—

1. Convergence of the condensate : $\lim C_p = 0$ for $T = 0$.
2. Convergence of the saturated vapour : $\lim C_p = \tfrac{5}{2} R$ for $T = 0$.

Cederberg puts for the constant of integration C

$$C = \log_{10} \pi_0,$$

whereas I had given the approximations (cf. p. 135),

$$C \sim 1\cdot1a \sim 0\cdot14 \frac{\lambda}{T_0} \qquad . \qquad . \qquad . \qquad (90)$$

Cederberg appears to ascribe (cf. p. 25) a profound theoretical importance to this hypothesis, for he calls the relation " äusserst wichtig " ; it may be shown, however, from dimensional considerations that Cederberg's rule (notwithstanding the practical utility which it may have) cannot possibly have any foundation in fact.

If we simplify Cederberg's vapour-pressure formula for low temperatures, we have

$$\log_{10} \frac{p}{\pi_0 T^{2\cdot5}} = - \frac{\lambda_0}{4\cdot57 T}.$$

On the right side of the equation there is a pure, i.e. dimensionless number $\left[\dfrac{\text{energy}}{\text{energy}}\right]$: the left side must therefore also be independent of the system of units employed. This is only possible if there is a dimensionless quantity after the logarithmic sign : the fact that Cederberg's rule does not comply with this requirement deprives it of any deep theoretical meaning. The exact calculation of C, which we shall learn in Chapter XIII, of course satisfies this requirement.

It is obvious, too, that nothing more than empirical regularities are to be looked for in the two relations **(90)** found empirically by me, nor have I ever looked for them. My criticism, I may repeat, is directed solely against the theoretical foundation, and in no way against the utility of Cederberg's rule, though even of this I may have certain doubts. In the calculation of vapour-pressure curves, we shall mostly have to forego the use of rigid theoretical formulæ for a long time, perhaps for ever; empirical regularities retain, therefore, a great importance, so that it appears worth while to investigate both Cederberg's and my vapour-pressure formulæ further, and to compare their utility. It may thus be possible, by further modification, to arrive at still better formulæ.

The calculation of chemical affinities may be made, of course, employing Cederberg's formula, in a manner entirely analogous to that in which it was performed above, employing mine; it will suffice, therefore, to refer once more to the very interesting results at which Cederberg has arrived in his important study.

CHAPTER XII

SOME SPECIAL APPLICATIONS OF THE HEAT THEOREM AND OF THE APPROXIMATION FORMULA DERIVED FROM IT

IN the foregoing chapter we have dealt with a number of applications of our formulæ rather from the point of view of the systematic justification of them; we shall now consider a number of further examples which have a certain interest of themselves, and will only incidentally supply further material for the consolidation of our ideas.

1. Determination of Thermo-chemical Data by Application of the Heat Theorem to Condensed Systems.—The fundamental equation of the Second Law,

$$A - U = T\frac{dA}{dT}$$

has been repeatedly employed for the estimation of thermo-chemical quantities, particularly in the form which it takes when applied to galvanic cells,

$$EF - U = TF\frac{dE}{dT};$$

but $\frac{dA}{dT}$, or, in the latter special case, the temperature coefficient of the galvanic cell, must be determined with great accuracy. This can be done to a limited extent only.

Now, according to the First Law,

$$U = U_0 + \Sigma\nu\int_0^T C_p dT = U_0 + E \quad . \quad (99)$$

and according to the New Heat Theorem,

$$A = U_0 - T\int_0^T \frac{E}{T^2} dT \quad . \quad . \quad . \quad (100)$$

We see then that we may determine first U_0, and then also U for all temperatures if—

1. A is known at a single given temperature; and
2. C_p is given down to the lowest temperatures for each body taking part in the reaction.

We have already (p. 115) had an example of such an application; the values found by U. Fischer for the heat of formation of silver iodide by means of formulæ **(99)** and **(100)** should be more reliable than that obtained by equation **(1)**. For though the electro-motive force of the cell in question could be determined at one temperature with sufficient accuracy (about 1 per 1000), the temperature coefficient could not be obtained so precisely as to allow the determination of U also to 1 per 1000 by equation **(1)**; for the use of formula **(100)**, on the other hand, the specific heats need be known with only moderate accuracy, since they serve only to determine the relatively small difference A − U. The case is similar in the following example.

In calculating the electro-motive force of the cell

$$Hg/HgCl/PbCl_2/Pb,$$

Pollitzer found a difference which led him to suppose that, since the heat of formation of calomel is very reliable, Thomsen's value for the heat of formation of lead chloride (82,700) must be several thousand calories too small. Thereupon Koref and Braune (98b) redetermined the value, using a very accurate calorimetric method, and found, as a matter of fact, the figure 85,570, which is higher by 2870. Pollitzer's suggestion, put forward on what were certainly very sound grounds, was therefore completely justified, and the criticisms directed against it by E. Cohen have proved to be unfounded (cf. paper 98b, p. 196).

SOME SPECIAL APPLICATIONS 157

2. Use of the Heat Theorem to Control Experimental Work.—The Second Law has long been used to subject to an independent test the accuracy of the measurements made in investigating chemical equilibria. The new Heat Theorem is frequently suited, perhaps even more highly, to this purpose, a fact which has not been sufficiently noted hitherto. We shall, therefore, show by means of a few examples how the Theorem has already led, or could have led, to the discovery of inaccuracies or of incorrect interpretation of observations.

In the ammonia equilibrium I found, (1) and (8), a large difference between the value calculated by means of the approximation formula and the measurements carried out by Haber and van Oordt, which, though stated to be preliminary, appeared to claim at least a certain degree of reliability. The redetermination of the constants of the equilibrium which was then made (cf. Jost, 13) showed that the earlier value was considerably in error, and confirmed satisfactorily that calculated by means of the approximation formula.

The determination of the equilibrium of the reaction $C + CO_2 = 2CO$, by Clement,* led to values which differed both from the older measurements of Boudouard and from those calculated by means of the approximation formula; a later paper by Rhead and Wheeler † confirms the latter values, and renders it probable that Clement's are wrong by a systematic error.

The case of the two sulphides of copper may also be mentioned. The dissociation pressure of sulphur above a mixture of the two crystalline sulphides should, from a calculation with the approximation formula, be equal to the vapour pressure of solid sulphur at the same temperature, if Thomsen's heats of formation are taken as a basis for the calculation. In contradiction to this, Frl. Wasjuchnow

---

* University of Illinois, No. 30 (1909).
† " Journ. Chem. Soc.," **97**, 2178 (1910).

(21) found that the sulphides must be warmed about 150° higher to obtain equality of pressure. The heats of formation were therefore subjected to a fresh determination, and H. v. Wartenberg (29a) found as a result that the figures of Thomsen were in error by 1300 and 1450 cals. respectively.

From Thomsen's determinations of the heat of formation of silver iodide it may be calculated, by means of the approximation formula, that iodine should have a measurable dissociation pressure at moderately high temperatures. Naumann (11) found, however, that this was too small to be measured even at 600°. Analogous differences occurred in the theoretical calculation of the E.M.F. of the silver-iodine electrode. U. Fischer then found the heat of formation by three independent methods as 15,200, 14,800, and 15,000, whereas Thomsen had given 13,800; so the differences between theory and observation are reconciled (cf. further, p. 113).

Thus in all these cases, in spite of the contradictions originally existing, the Theorem and even the approximation formula have proved themselves reliable and have led to a correction or revision of the observations.

3. Electro-chemical Applications.—If we calculate with A in gm.-cals. we have for the E.M.F.

$$E = \frac{A}{n \times 23046} \text{ volts,}$$

where n denotes the number of electro-chemical equivalents involved in the reaction considered.

If we are concerned with cells in which pure substances only occur, the calculation is at once clear, as we have seen already (p. 113 *et seq.*) from several examples; in galvanic arrangements which are not composed of pure substances alone it is not difficult, in the majority of cases, to recalculate them to this condition. Let us consider as an example the cell

$Pb/PbBr_2$ saturated solution/Br_2/Pt.

SOME SPECIAL APPLICATIONS

The process supplying the current is given by the equation
$$Pb + Br_2 = PbBr_2.$$
The substances Pb and $PbBr_2$ are here in the pure state, but the liquid bromine dissolves some water. According to the law of relative lowering of solubility, the solubility is depressed by the water taken up, and the E.M.F. is correspondingly somewhat reduced. The solubility of bromine, and thus also the E.M.F., may be recalculated by means of the above-mentioned law to the state of complete purity.

When gas cells are concerned we have to make use of formulæ **(75)** and **(77)**, and then to calculate the E.M.F. according to well-known rules.

As an example, let us deduce the E.M.F. of the oxyhydrogen cell. If hydrogen and oxygen are present in the cell under atmospheric pressure, we have *

$$E = \frac{0 \cdot 0001983 T}{4} \log \frac{1}{K' \pi^2},$$

where π is the value of the vapour pressure of water in atmospheres (= 0·0191 at T = 290) and K′ is to be calculated as on page 22. It follows (cf. p. 23) that

$$E = 1 \cdot 2325 \text{ volts at } T = 290.$$

If we make the calculation with the approximation formula **(92)**, page 137, we have

$$\log K' = -\frac{115160}{4 \cdot 571 T} + 1 \cdot 75 T - 1 \cdot 2,$$

which gives
$$E = 1 \cdot 25 \text{ volts,}$$

a much better value than was given by the direct measurements (1·15 volts, cf. p. 23). At higher temperatures, where the influence of the specific heats becomes more and more marked, the approximation formula would, of course, give values appreciably in error.

* " Theor. Chem.," p. 849.

In general, the above simply manipulated approximation formula appears to be very useful for obtaining a first rough idea of the calculated electromotive forces at ordinary temperatures. Bodländer has thus compared the heats of formation of various iodides and chlorides with the electromotive forces of the respective cells. It was then found that the differences between the two values were only slight in the case of the iodides; according to our approximation formulæ

$$23046E = U_0 - \beta T^2, \quad U = U_0 + \beta T^2,$$

this is because the influence of the coefficient β is only small, and it has different signs for different combinations. In the case of the chlorides, however, Bodländer obtained a good agreement, on the average, when he subtracted

$$5060 \text{ cals.} = 0\cdot 22 \text{ volts}$$

from the heats of formation per equivalent. The above treatment shows that, neglecting again the coefficient β, which is always small, the difference between the two quantities amounts to

$$\frac{4\cdot 571 \times 290(1\cdot 75 \log T + 3\cdot 2)}{2} = 4971 \text{ cals.} = 0\cdot 217 \text{ volts.}$$

Bodländer's empirical result is therefore explained quantitatively by our theory, just as on page 149 the empirical coefficient of the le Chatelier-Matignon rule received a theoretical interpretation.

Lastly, it is an illuminating fact that just as gaseous equilibria may be calculated from thermal quantities and vapour pressures (or chemical constants), so also electrode potentials may be determined from thermal quantities and solubilities.

For this purpose, let us consider simply some particular example, say the electrode Ag/I_2; if, then, P_1 and P_2 denote the solution tensions of the two electrodes, and p_0 be the osmotic pressure of an aqueous solution saturated with silver

SOME SPECIAL APPLICATIONS

iodide, we find the electromotive force, first according to the osmotic and then according to the thermodynamic theory:—

$$RT \log_e \frac{P_1}{p_0} + RT \log_e \frac{P_2}{p_0} = \epsilon_1 - \epsilon_2 - RT \log_e p_0^2$$

$$= \frac{U_0 - \beta T^2 - \frac{\gamma}{2} T^3}{23046} \quad . \quad . \quad . \quad (100)$$

We thus see that it is possible to calculate the solution tension characteristic of each electrode (except for a constant factor) or the electrode potentials (except for an additive constant). In the practical application of the figures the constant naturally disappears since the electrode potential may be arbitrarily assumed for one electrode, e g. equated to zero for the hydrogen electrode. Now the osmotic theory allows the E.M.F. of any galvanic arrangement to be determined in which dilute aqueous solutions are employed; we thus see that this theory is amplified by our thermodynamic treatment, in that the electrode potentials are now made accessible to simple theoretical calculations, instead of having to be determined for a given temperature, when using the osmotic theory, by means of measurements carried out on each of the electrodes in question; for the electrode potentials are derivable from thermal data and solubilities.

For each electrode potential ϵ there is obviously needed only one difficultly-soluble salt; the solubilities of all the other difficultly-soluble salts are then calculable from the corresponding thermal data.

It will be a problem for future work to calculate out the existing experimental material from this point of view, and to obtain fresh data.

4. Application to Photochemical Side-reactions.—Einstein has pointed out that the application of the quantum theory to photochemical processes renders probable the relation

$$N = \frac{Q}{h\nu};$$

here Q is the heat absorbed, ν the oscillation frequency of the light absorbed, and N the number of molecules split up by the light. Just as in electrolysis, the process which is actually observed is disturbed by subsidiary reactions; in order to test the above formula one must be able to take these into account.

As shown by Warburg, in a very important paper,[*] this may be done, in some circumstances, by applying the approximation formula. He examined the photolysis of hydrobromic acid gas by ultra-violet light, which can only take place according to the equation

$$HBr = H + Br;$$

the question arises whether hydrogen or bromine atoms are capable of acting on hydrogen bromide.

We put (in round figures)—

$H + H = H_2$ . . . $+ 100{,}000$ cals.
$Br + Br = Br_2$ . . . $+ 46{,}000$,,
$H_2 + Br_2 = 2HBr$ . . $+ 24{,}000$,,

It is then easily calculated that—

(a) $H + Br = HBr$ . . . $+ 85{,}000$ cals.
(b) $Br + HBr = Br_2 + H$. $- 39{,}000$,,
(c) $H + HBr = H_2 + Br$. $+ 15{,}000$,,

We take as chemical constants for H and H_2 1·6, and for the other molecules, 3·2; then, expressing the concentration of H by [H], and so on,

$$\log \frac{[Br][HBr]}{[Br_2][H]} = + \frac{39000}{4 \cdot 57T} + 1 \cdot 6$$

$$\log \frac{[H][HBr]}{[H_2][Br]} = - \frac{15000}{4 \cdot 57T}.$$

For T = 290 (room temperature) the left side takes the values $+ 31 \cdot 1$ and $- 11 \cdot 3$ respectively in the above two cases.

[*] "Ber. Berl. Akad.," 24 Feb., 1916.

SOME SPECIAL APPLICATIONS

Bromine atoms therefore do not act appreciably on hydrogen bromide, but hydrogen atoms must react with HBr molecules to form hydrogen molecules and bromine atoms.

The free bromine atoms, of course, rapidly disappear according to the equation

$$2Br = Br_2 ;$$

we thus obtain the result, already found by Warburg, that the primary yield of bromine is doubled by the subsidiary reaction, and on this assumption Einstein's law is shown to be fulfilled.

From the purely chemical aspect, it is at first glance a very surprising result that hydrogen atoms ("nascent hydrogen") should oxidize hydrogen bromide to free bromine.

The corresponding calculation applied to the photolysis of the hydrogen-chlorine mixture, leads to similarly remarkable results. Here the primary photo-chemical process is obviously to be regarded as that of the reaction,

$$Cl_2 = 2Cl.$$

At any rate, no other hypothesis will be made so long as this simple assumption suffices. We have in this case

$$Cl + Cl = Cl_2 \quad . \quad . \quad . \quad + 106{,}000^* \text{ cals.}$$
$$H + H = H_2 \quad . \quad . \quad . \quad + 100{,}000 \text{ ,,}$$
$$H_2 + Cl_2 = 2HCl \quad . \quad . \quad . \quad + 44{,}000 \text{ ,,}$$

Hence—

$$(d)\ H + Cl = HCl \quad . \quad . \quad . \quad + 125{,}000 \text{ cals.}$$
$$(e)\ Cl + H_2 = HCl + H \quad . \quad . \quad + 25{,}000 \text{ ,,}$$
$$(f)\ H + Cl_2 = HCl + Cl \quad . \quad . \quad + 19{,}000 \text{ ,,}$$

Application of the approximation formula to (e) and (f) shows at once that chlorine atoms plus hydrogen molecules give free hydrogen atoms with formation of hydrogen

* Cf. p. 148.

chloride; the hydrogen atoms, in turn, plus chlorine molecules, regenerate free chlorine atoms, again with formation of hydrogen chloride. If, therefore, chlorine molecules are split up by the light, a very much greater yield of hydrogen chloride is to be expected than corresponds with the primary production of chlorine atoms, i.e. with Einstein's law. It is now known that this is actually the case, the yields being more than a million times greater.

Of course, both hydrogen atoms and chlorine atoms disappear at the same time, according to the three equations:—

$$H + H = H_2,$$
$$Cl + Cl = Cl_2,$$
$$H + Cl = HCl,$$

so that there is not a quantitative formation of hydrogen chloride when the number of primary chlorine atoms formed is small. The yield of HCl must also decrease if there are impurities present, which remove chlorine or hydrogen atoms; as a matter of fact, the photolysis of the chlorine-hydrogen mixture is very sensitive towards certain impurities.

As we saw above, the reaction

$$Br + H_2 = HBr + H$$

does not occur to an appreciable extent. If, therefore, again on the simplest assumption, the primary action of light on a mixture of hydrogen and bromine vapour consists in the splitting up of bromine molecules, this cannot lead to the formation of appreciable quantities of hydrogen bromide. This corresponds with the known fact that the mixture is apparently almost completely insensitive to light, as has quite recently been confirmed by special experiments (not yet published) by Frl. Dr. Pusch.

The further employment of the method introduced by Warburg promises a wealth of results in connection not only with photo-chemistry, but also with pure chemistry, particularly as regards atomic reactions.

SOME SPECIAL APPLICATIONS

5. Other Applications.—It is not the object of this book to deal with all the applications of the Heat Theorem or of the approximation formula derived from it ; we shall only refer here to the following additional investigations emanating from our laboratory : v. Wartenberg (6), Nernst (8), Preuner (9), v. Wartenberg (10), Bodenstein and Dunant (12), Halla (14), Nernst (19), Wasjuchnowa (26), Nernst (25), Pollitzer (26), Horak (29), Bodenstein and Katayama (29b), Koref (31), Nernst (33), Nernst (42), Holland (45), Halla (50), v. Wartenberg (69), Holland (70), Wolff (90), and v. Kohner (100).

CHAPTER XIII

THEORETICAL CALCULATION OF CHEMICAL CONSTANTS

1. Statement of the Problem.—For monatomic gases the following vapour-pressure formula holds (cf. p. 124):—

$$\log_e p = -\frac{\lambda_0}{RT} + 2\cdot 5 \log_e T + i, \quad . \quad . \quad (101)$$

and the heat of evaporation λ at the temperature T has the value

$$\lambda = \lambda_0 + C_p T = \lambda_0 + 2\cdot 5 RT \quad . \quad . \quad (102)$$

A region of temperature is here presupposed so low that the specific heat of the condensate may be neglected. This supposition entails no limitation, serving merely as a simplification; if it is not fulfilled, there merely appears a further term in formula **(102)** which, of course, results in a corresponding further term in formula **(101)**.

Supplementing the classical thermodynamics, which has nothing to say on the point, our Heat Theorem shows that the constant of integration i depends upon the nature of the gas concerned, and not on the type of the physical or chemical equilibrium.

It is clear that the significance of i is thus vastly enhanced and the idea arises that i might be capable of being calculated simply from theory. Many workers have been prompted to take up this question, and an extremely important and remarkable result has been obtained, the relation for i being found to be

$$i = \log_e \frac{(2\pi m)^{3/2} k^{5/2}}{h^3}; \quad . \quad . \quad (103)$$

CALCULATION OF CHEMICAL CONSTANTS

here $m = \dfrac{M}{N}$ is the mass of the atom, $k = \dfrac{R}{N}$ (N = number of molecules per gramme molecule), and h = Planck's constant.*
Equation **(103)** may be brought into the form :—

$$i = \log_e \frac{(2\pi)^{3/2} k^{5/2}}{N^{3/2} h^3} + 1 \cdot 5 \, \log_e M = i_0 + 1 \cdot 5 \, \log_e M, \quad (104)$$

where i_0 is thus a constant independent even of the nature of the gas.

Before we deal with these theories we shall examine the treatment of di- and poly-atomic gases.

For rigid, i.e. sufficiently cooled diatomic gaseous molecules, we have the equations,

$$C_v = \tfrac{5}{2}R, \quad C_p = \tfrac{7}{2}R; \qquad \qquad (105)$$
$$\lambda = \lambda_0' + 3 \cdot 5 \, RT; \qquad \qquad (106)$$

$$\log_e p = -\frac{\lambda_0'}{RT} + 3 \cdot 5 \, \log_e T + i''. \qquad (107)$$

Now we know (p. 74) that with diatomic gases the energy of rotation is lost when the cooling is sufficient, i.e. we have then, as for monatomic gases,

$$C_v = \tfrac{3}{2}R, \quad C_p = \tfrac{5}{2}R. \qquad \qquad (108)$$

At these temperatures, therefore, the same formulæ hold as for monatomic gases, namely, **(102)** and **(101)**, instead of **(106)** and **(107)**; it remains to establish the relation between λ_0 and λ_0', and between i and i''.

Let us put, in general, for a diatomic gas,

$$C_v = \tfrac{3}{2}R + \frac{dE}{dT}, \quad C_p = \tfrac{5}{2}R + \frac{dE}{dT}. \qquad (109)$$

where E is to represent the amount of energy of rotation; these equations are valid, of course, only up to temperatures

---

* For numerical values see Appendix.

at which the energy of the internal vibrations may be neglected. Then,

$$\lambda = \lambda_0 + \tfrac{5}{2}RT + E \quad . \quad . \quad . \quad (110)$$

$$\log_e p = -\frac{\lambda_0}{RT} + 2\cdot 5 \log_e T + \frac{1}{R}\int_0^T \frac{E}{T^2} dT + i \quad (111)$$

At higher temperatures these two equations must reduce to equations (106) and (107); we shall suppose that at T' equations (110) and (111) have become practically identical with (106) and (107), so that we may put

$$\lambda = \lambda_0 + \tfrac{5}{2}RT + E_1 + R(T - T') = \lambda_0 + \tfrac{7}{2}RT + E_1 - RT' \quad (112)$$

where E_1 corresponds with the temperature T'.

If we wish to integrate the equation

$$\lambda = RT^2 \frac{d \log_e p}{dT} \quad . \quad . \quad . \quad (112a)$$

with the aid of (112), we must note that the latter holds only above T'; below T' we have the relation

$$\lambda = \lambda_0 + \tfrac{5}{2}RT + E.$$

We thus obtain

$$\log_e p = -\frac{\lambda_0}{RT} + 2\cdot 5 \log_e T + \frac{1}{R}\int_0^{T'} \frac{E}{T^2} dT \\ + \frac{1}{R}\int_{T'}^{T} \frac{E_1 + R(T-T')}{T^2} dT + i$$

or, transposing,

$$\log_e p = -\frac{\lambda_0 + E_1 - RT'}{RT} + 3\cdot 5 \log_e T + \frac{1}{R}\int_0^{T'} \frac{E}{T^2} dT \\ + \frac{E_1}{RT'} - \log_e T' - 1 + i \quad . \quad . \quad . \quad (113)$$

Comparison of (113) and (107) shows that

$$\lambda_0' = \lambda_0 + E_1 - RT',$$

which may also be deduced from a direct comparison of **(112)** and **(106)**; the constant of integration i' of formula **(107)** comes out to

$$i' = \frac{1}{R}\int_0^{T'}\frac{E}{T^2}dT + \frac{E_1}{RT'} - \log_e T' - 1 + i. \qquad (114)$$

This expression is of course independent of T', since this temperature has been chosen sufficiently high, i.e. in the region where the normal value $C_v = \frac{5}{2}R$ holds for diatomic gases.

The somewhat abstract character of the above calculations is more readily grasped if a definite function is used in making them; for example, the increase in the energy of rotation with the temperature may be expressed by means of an Einstein function, which should at least come very near to the truth; in this case a simple expression may be obtained for the relation between i and i'. We shall frequently have to carry out such calculations.

The conditions in the case of tri- and poly-atomic gases are sufficiently analogous that we shall not deal further with them here; if the law for the increase of rotational energy is known, the i' of the equation

$$\log_e p = -\frac{\lambda_0'}{RT} + 4\log_e T + i' \qquad . \qquad (115)$$

may again be deduced from the corresponding value of i in equation **(101)**.

Thus we recognize how wide is the scope of equation **(101)**, and how great the importance which consequently attaches to a determination of i by theoretical means.

2. Theoretical Calculation of the Integration Constant of the Vapour-pressure Formula.—Equation **(103)** was obtained almost simultaneously by Sackur * and by Tetrode †; we

---

* "Nernst-Festschrift," p. 405, 1912; "Ann. d. Phys.," **40**, 67, 1913.

† "Ann. d. Phys.," **38**, 434; **39**, 255, 1912.

can do no more than refer to the extremely original methods of treatment used by these investigators, but we shall briefly explain a method devised by O. Stern,* which also leads to the same equation.

The idea underlying Stern's calculation is as follows: Our Heat Theorem shows that i is independent of the nature of the condensate, and even of the nature of the equilibrium concerned. Thus, if it is possible to find a model, as simple as possible, of a condensate for which we can calculate the equilibrium with its saturated vapour, there is a very great probability that the value of i so obtained should be in fact that which also applies to all other cases, however complicated they may be.

This probability would become a certainty if we suppose the model in question to be of natural construction, i.e. if it really has the essential properties of a condensate, e.g. of a crystal ; Stern's model is, however, somewhat far removed from this desideratum. But the probability will still be almost a certainty if reliable experimental proof can be adduced, in some cases at least, that the value of i in Stern's model recurs in equilibria which are actually measured ; this it always does.

I now give, with a few slight modifications, Stern's description and basis of his model, and also the derivation of the condition for equilibrium.

In order to derive a vapour-pressure formula kinetically, a model of molecular mechanism must be constructed which illustrates the solid in equilibrium with its vapour. Any such model must, if it is to conform with the equations of mechanics, show the same thermodynamically calculable dependence of vapour pressure on temperature for the same ν, λ_0, and molecular weight. The representation here used is as follows: Let there be, in a given space, points P which attract the atoms with a force directly proportional to the distance r. Since the heat of evaporation has a

* "Physik. Zeitschr.," **14,** 629, 1913.

CALCULATION OF CHEMICAL CONSTANTS

finite value this force acts only up to a certain distance s. The points P are therefore surrounded by spherical fields of attraction within which the atoms oscillate as monochromatic resonators, while in the remainder of the space they move about as ideal gas molecules, free from constraint. The potential energy of the molecule at any point in space is thus determined and, further, the ratio of molecular density in the spheres to that in the free space is given by Boltzmann's e-Law. Since it is only the ratio of the densities that is fixed, we must make a further special hypothesis as to the molecular density prevailing in the spheres in order to determine the vapour density. It is scarcely possible to make any other assumption than that there is on the average one molecule in each sphere. For we do suppose that in real solids there are the same number of atoms as positions of equilibrium (points P). The vapour density is now completely determined, and the calculation takes the following form. If we denote the molecular density and potential energy of a molecule in the free space by n_0 and ψ_0, and those at a point in a sphere distant r from the equilibrium position P by n_r and ψ_r, we have, according to Boltzmann, the equation

$$n_r : n_0 = e^{-\frac{\psi_r}{kT}} : e^{-\frac{\psi_0}{kT}}$$

or,

$$n_r = n_0\, e^{\frac{\psi_0}{kT}} \cdot e^{-\frac{\psi_r}{kT}}$$

The total number of molecules contained in a sphere is, by hypothesis,

$$n = \int_0^s n_r \cdot 4\pi r^2 dr = \int_0^s n_0 \cdot e^{\frac{\psi_0}{kT}} \cdot e^{-\frac{\psi_r}{kT}} 4\pi r^2 dr = 1.$$

If now the force with which the molecule (of mass m) is attracted by the centre of the sphere is determined by the equation

$$m\frac{d^2r}{dt^2} = -a^2 r,$$

the potential energy ψ_r of the molecule at a distance r is $\dfrac{a^2}{2}r^2$, and that in the free space, ψ_0, is equal to $\dfrac{a^2}{2}s^2$. We therefore have the equation

$$\mathrm{I} = n_0 e^{\frac{\psi_0}{k\mathrm{T}}} 4\pi \int_0^s e^{-\frac{a^2}{2}r^2 \big/ k\mathrm{T}} r^2 dr.$$

If we put $\dfrac{\frac{a^2}{2}r^2}{k\mathrm{T}} = x^2$ and $\dfrac{\frac{a^2}{2}s^2}{k\mathrm{T}} = x_0^2$, then

$$\mathrm{I} = n_0 \cdot e^{\frac{\psi_0}{k\mathrm{T}}} 4\pi \left(\frac{k\mathrm{T}}{\frac{a^2}{2}}\right)^{3/2} \int_0^{x_0} e^{-x^2} x^2 dx.$$

Thus

$$p = n_0 k\mathrm{T} = \frac{\left(\frac{a^2}{2}\right)^{3/2}}{4\pi(k\mathrm{T})^{1/2}} \cdot e^{-\frac{\psi_0}{k\mathrm{T}}} \cdot \frac{1}{\int_0^{x_0} e^{-x^2} x^2 dx} \qquad . \quad \textbf{(116)}$$

Formula **(116)** thus allows the vapour pressure to be calculated from the properties of the model used. The further treatment of the question is now very simple.

In equation **(116)** we may first put

$$e^{-\frac{\psi_0}{k\mathrm{T}}} = e^{-\frac{N\psi_0}{R\mathrm{T}}} = e^{-\frac{\lambda_0'}{R\mathrm{T}}}$$

and

$$a^2 = m(2\pi\nu)^2.$$

We then have to evaluate the integral, to do which we make use of the well-known recurrence formula,*

$$\int e^{-x^2} dx = xe^{-x^2} + 2\int x^2 e^{-x^2} dx.$$

Now, in the region of low vapour pressures which is here concerned,

$$x_0^2 = \frac{\lambda_0'}{R\mathrm{T}}$$

* Cf., for example, Nernst and Schönfliess, " Mathem. Behandl. d. Naturw.," 7 Aufl., p. 353.

CALCULATION OF CHEMICAL CONSTANTS

is large compared with 1 for all known solids, so that e^{-x_0} is very small compared with 1. The relation above found for the model with which we are dealing,

$$\frac{n_r}{n_0} = \frac{e^{\frac{\psi_0}{kT}}}{e^{\frac{\psi_r}{kT}}},$$

also shows this, since $\frac{n_r}{n_0}$ must be a large number while $e^{\frac{\psi_r}{kT}}$ cannot be less than 1, however small ψ_r may be. We may therefore put

$$\int_0^{x_0} e^{-x^2} x^2 dx = -\tfrac{1}{2}[xe^{-x^2}]_0^{\infty} + \tfrac{1}{2}\int_0^{\infty} e^{-x^2} dx = \frac{\sqrt{\pi}}{4}.$$

After substituting the above relations and introducing logarithms, formula **(116)** reduces to

$$\log_e p = -\frac{\lambda_0'}{RT} - 0\cdot 5 \log_e T + \log_e \frac{(2\pi m)^{3/2} \nu^3}{k^{1/2}} \quad \textbf{(117)}$$

It must be remembered that the foregoing vapour-pressure formula depends on the classical kinetic theory, and that none of the laws of the quantum theory have been used in its derivation. It holds therefore for a solid which strictly obeys the law of Dulong and Petit ($C_v = 3R$) and for which there is no difference between C_p and C_v (cf. p. 65), i.e. which shows no thermal expansion.

Equation **(117)** might therefore have been obtained before the quantum theory had been put forward, but, although very remarkable, it could hardly have been used at all in practice because, until our Heat Theorem was propounded, there was no reason to ascribe a general significance to the last term of this equation.

We now apply the quantum theory to solids so far as to conclude that C_v decreases at low temperatures, and does so, since it consists of purely monochromatic resonators of the frequency ν, according to Einstein's formula (p. 59).

Classical thermodynamics then gives us the relations

$$\lambda = \lambda_0 + 2\cdot 5 RT - 3R\frac{\beta v}{e^{\frac{\beta v}{T}} - 1}, \qquad (118)$$

$$\log_e p = -\frac{\lambda_0}{RT} + 2\cdot 5 \log_e T + 3 \log_e (e^{\frac{\beta v}{T}} - 1) \\ - 3\frac{\beta v}{T} + i; \quad (119)$$

the latter equation is obtained by integration of equation **(112a)**, making use of formulæ **(51)** to **(53)** of Chapter VIII.

If we proceed to higher temperatures, equation **(119)** must reduce to **(117)**; actually, if we note that for high temperatures $\log_e (e^{\frac{\beta v}{T}} - 1)$ reduces to $\log_e \frac{\beta v}{T}$, **(119)** takes the form,

$$\log_e p = -\frac{\lambda_0 + 3R\beta v}{RT} - 0\cdot 5 \log_e T + 3 \log_e \beta v + i \quad (120)$$

Equations **(120)** and **(117)** thus become identical if we put

$$\lambda_0' = \lambda_0 + 3R\beta v$$

$$3 \log_e \beta v + i = \log_e \frac{(2\pi m)^{3/2} v^3}{k^{1/2}}.$$

The latter relation gives, however, $\left(\beta = \frac{h}{k}\right)$

$$i = \log_e \frac{(2\pi m)^{3/2} k^{5/2}}{h^3} \qquad . \qquad . \quad (103)$$

as already indicated above (p. 166).

We shall not deal here with certain criticisms of the above derivation which (in contrast with Sackur's theory) applies no quantum considerations of any sort to the gas itself; they are met in part by an analogous but very generalized derivation published by Tetrode [*] in 1915. Stern's argument finds considerable support from its final result, formula **(103)**, which is of extreme interest, and in

---

[*] "Akad., Amsterdam," 27 Feb., 1915.

CALCULATION OF CHEMICAL CONSTANTS

which there appear none of the peculiarities of the model assumed ; further support is, of course, given by the almost convincing confirmation which it receives from experiment, with which we shall now deal.

3. Experimental Test.—Calculations of a somewhat provisional character to check formula **(103)** have been already put forward by Sackur (*loc. cit.*) and by Tetrode (*loc. cit.*) ; in view of the great importance of the whole question, I have occupied myself, with the assistance of my co-workers, in extending and critically reviewing the available material.

For the purpose of testing formula **(103)**, only such equilibria come into consideration at present as relate to monatomic gases, or to gases which it has been possible to convert by cooling into the " thermally-monatomic " state ; this has only been possible hitherto with hydrogen (p. 71).

In addition to hydrogen, there is at present adequate experimental material only for mercury and argon.

Mercury.—Let us consider, as the reaction, the evaporation of liquid and solid mercury, the atomic heats for which are well known. According to Kurbatoff,* the heat of evaporation at 358° amounts to 13,600 per mol. On the assumption that the gas laws hold for mercury up to the boiling-point, and making use of the mean specific heat of liquid mercury determined by the same observer, the heat of evaporation of liquid mercury at 21° works out to

$$\lambda_{21°} = 13{,}600 + 2528 - 337 \times 4 \cdot 963 = 14{,}455.$$

We shall see, however, that this value is some 2 per cent. too low. The deviation of saturated mercury vapour from the ideal gaseous state reduces its specific heat (cf. *infra*), and therefore increases $\lambda_{21°}$, on a rough estimate, by some 80 cals. Although the error still remaining appears inconsiderable in view of the difficulty of the measurement, it is sufficient to throw out the result we are calculating by a large amount.

* " Zeitschr. f. physik. Chemie," **43**, 107, 1903.

Fortunately, there are a number of good vapour-pressure measurements available for mercury which, combined together, give us the heat of evaporation with an accuracy of about 1 pro mille. If we limit ourselves to pressures not exceeding a few tenths of an atmosphere we may use the equation of Clausius and Clapeyron,

$$\lambda = T\frac{dp}{dT}(v - v')$$

in the form *

$$\lambda = 4 \cdot 571 T^2 \frac{d \log_{10} p}{dT} = 4 \cdot 571 \frac{T_1 T_2}{T_2 - T_1} \log_{10} \frac{p_1}{p_2}.$$

The atomic heat of the vapour at constant pressure is 4·963, and we may regard that of liquid mercury † as nearly enough constant over the interval $-34°$ to $+260°$ and call it 6·64. For our purpose, therefore,

$$\lambda = \lambda_0 - 1 \cdot 68 T$$

holds with sufficient accuracy and hence, as had already been found by Hertz,

$$\log_{10} p = -\frac{A}{T} - 0 \cdot 847 \log T + B; \qquad \textbf{(120}a\textbf{)}$$

this equation is valid, of course, only in the above-mentioned range of temperature. It may be mentioned, merely for the sake of completeness, that B and the constant of integration which we require have nothing to do with one another.

For

$$A = \frac{\lambda_0}{4 \cdot 571},$$

Hertz derived the value 3342 from his measurements over the range 90°-200°, while Knudsen finds between 0° and 50°

---

* Cf., for example, Nernst, " Theor. Chem.," p. 66; $T_2 - T_1$ must, of course, be small compared with T_1 and T_2.

† Cf. the collation in Landolt-Börnstein's " Tabellen," p. 761 (1912): the increase found by Pollitzer just above the melting-point can have only an inappreciable influence on the result.

the identical figure 3342·26. Knudsen * makes the integration constant B slightly smaller ; this means that either the measurements of Hertz are on the average 2 to 3 per cent. too high, or that those of Knudsen are too low by a similar amount ; the latter is the less probable. The heat of evaporation λ_0, the value of which depends only on the ratio of the pressures at different temperatures, has an identical value assigned to it by both observers.

A really good value for the heat of evaporation is also given by the series of observations due to Pfaundler (cf. Knudsen, *loc. cit.*), while Morley's figures (cf. *ibid.*) show a very different trend : the latter, in view of their considerable deviation from those of Knudsen and Hertz, cannot apparently be taken into consideration in deciding on the heat of evaporation.

That Knudsen's values, in particular, are very reliable is demonstrated by the following test. In calculating out the figures obtained by Ramsay and Young for temperatures above 220° which are, as we know, very good, Knudsen finds that his formula gives pressures higher by 1 to 2 per cent. in the interval from 220° to 280°. If, therefore, Knudsen's numbers for 0° are combined with those of Ramsay and Young in the above-mentioned range of temperature, where we may assume the gas laws to hold rigidly (the pressures concerned are less than 0·2 atm.), the heat of evaporation would only come out 1 to 2 p. mille smaller. Now A. Gebhardt's review † shows that in this range of temperature both his later measurements and those of Jewett give numbers a few per cent. higher than the measurements of Ramsay and Young ; the most probable mean is therefore in complete agreement with Knudsen's measurements.

In these circumstances the heat of evaporation derived from the observations of Hertz and Knudsen must be

* " Ann. d. Physik," (4), **29**, 179, 1909.
† " Verhandl. d. Deutsch. physik. Ges.," 1905, p. 186.

credited with an accuracy of at least 1 pro mille. Hence

$$\lambda = 15{,}277 - 1{\cdot}68 T$$

(valid from $-32°$ to $+260°$).

According to this, the heat of evaporation at the melting-point ($T = 234{\cdot}4°$) amounts to 14,884, and that for solid mercury to $14{,}884 + 555 = 15{,}439$; the figure 555 represents the heat of fusion determined by Pollitzer, and confirmed, by a totally different method, by Koref.* For the absolute zero, the heat of evaporation of solid mercury is thus

$$\lambda_0 = 15{,}439 - 234{\cdot}4 \times 4{\cdot}963 + \int_0^{234{\cdot}4} c \, dT \quad . \quad (120b)$$

The atomic heat of solid mercury has been measured, by the two methods worked out in our laboratory, by Koref (*loc. cit.*), and then down to 31° abs. by Pollitzer (46); Koref's value fits perfectly into Pollitzer's series of observations. Pollitzer represented the measurements available by the formula of Lindemann and myself ($\beta \nu = 97$), plus a small additional term ($= C_p - C_v$). We now know that Debye's function is to be preferred, although the difference is very slight. It may easily be shown that when using this function one must put

$$c = f\left(\frac{96}{T}\right) + 21 \times 10^{-5} T^{3/2};$$

the value of $\beta \nu$ in the above formula can hardly be in error by as much as 1°.

Making use of the table † which I have calculated, the value of the integral in equation **(120b)** is found to be 1252, whence it follows that

$$\lambda_0 = 15{,}528.$$

* "Ann. d. Physik," (4), 56, 1911.
† "Ber. Berl. Akad.," 12 Dec., 1912; cf. also the Appendix.

CALCULATION OF CHEMICAL CONSTANTS

For the vapour pressure of monatomic gases we have in general (cf. **101**)

$$\log_e p = -\frac{\lambda_0}{RT} + 2{\cdot}5 \log_e T - \frac{1}{R}\int_0^T \frac{E}{T^2} dT + i \ . \quad (121)$$

or, after introducing common logarithms and substituting numerical coefficients,

$$\log_{10} p = -\frac{\lambda_0}{4{\cdot}571 T} + 2{\cdot}5 \log_{10} T - \frac{1}{4{\cdot}57}\int_0^T \frac{E}{T^2} dT + C \quad (122)$$

where

$$i = 2{\cdot}303 C. \quad (123)$$

If, in order to calculate C, we apply the above formula to the melting-point, p denotes the vapour pressure at the triple point; it follows from equation (**120**a), if we reckon p in atmospheres, that

$$\log_{10} p = -8{\cdot}574 \ (T = 234{\cdot}4°).$$

The integral of equation (**122**) comes out, using again the tables I have calculated, to

$$\frac{8{\cdot}186}{4{\cdot}57} + \frac{0{\cdot}188}{4{\cdot}57} = 1{\cdot}832,$$

and thus equation (**122**) becomes, in our case,

$$-8{\cdot}574 = -14{\cdot}493 + 5{\cdot}925 - 1{\cdot}832 + C,$$

or

$$C = 1{\cdot}83 \pm 0{\cdot}03.$$

As to the reliability of this figure, the value on the left-hand side of the equation above and the first figure on the right can hardly contain errors amounting to more than 0·01 to 0·02; the second figure on the right-hand side is free from error; as regards the third, an error in $\beta\nu$ of 1° would alter C by 0·01; in the aggregate no error greater than 0·03 can well be assumed.

It will probably be some time before there is another

case in which all the elements for the calculation are available in such perfection as here ; possibly the above discussion may help to make investigation on these lines more extensive.

Argon.—The heat of evaporation at the boiling-point, the heat of fusion, and the specific heat of solid argon down to $T = 17·8$ are given in a paper recently published by Eucken (105) ; the vapour pressures of solid and liquid argon have been determined by Travers, and recently, in greater detail, by Crommelin.*

The vapour pressures with which we shall have to deal are in the region of 0·3 to 1·2 atm. ; in view of the low temperature and of the resulting relatively high molar concentrations, we can no longer assume that the gas laws are sufficiently well obeyed.

In order to be able to apply the necessary corrections, we shall make use of the equation of state of Daniel Berthelot. † Though this may not perhaps be very accurate at temperatures which are much below the critical, we shall come considerably nearer the truth if we apply the requisite corrections (which are always only small) than if we simply calculate on the gas laws ; we shall, in any case, obtain an idea of the extent of the uncertainty caused by the deviation from the gas laws.

The atomic heat of liquid argon in the neighbourhood of the boiling-point was found by Eucken to be constant at 10·5, so that the temperature coefficient of the heat of evaporation would amount to $10·5 - 5·0 = 5·5$ cals. if the gas laws held for the gaseous argon. Actually, the atomic heat of the saturated argon vapour is reduced because, as the density of the vapour increases with the temperature, the well-known Joule effect for imperfect gases comes in. This

* " Communications Leiden," No. 138 (1913), and No. 140 (1914).
† Cf. " Theoret. Chemie," p. 257. A new test of this equation, for saturated vapours in particular, has just been carried out by Herr Schimank (104) at my suggestion, and he has established its applicability.

CALCULATION OF CHEMICAL CONSTANTS

is calculated, according to Berthelot, to be

$$cd\mathrm{T} = \frac{27}{32}\left(\frac{\theta_0}{\mathrm{T}}\right)^2 1.985 \frac{dp}{\pi_0}\theta_0 \qquad . \qquad . \quad (124)$$

The critical temperature $\theta_0 = 150.7$, the critical pressure $\pi_0 = 48.0$ atm. It may be calculated from the above equation that for the temperature range from 87.25° (boiling-point) to 83.79° (melting-point), where the vapour pressure increases by 0.1 atm. per degree, the saturated vapour of argon disposes of 1.6 cals. per degree per mol by reason of its expansion; hence the heat of evaporation must fall off by $5.5 + 1.6 = 7.1$ cals. per degree.

Eucken found for the boiling-point a heat of evaporation of 1500 cals.; at the melting-point it is therefore 1525 cals. The heat of fusion was found by Eucken to be 268 cals., so that we arrive at the heat of evaporation of solid argon at the melting-point as 1793 cals.

As we shall be specially interested in the heat of fusion later on, we shall, as a control, calculate its value by means of the formula of Ratnowsky,*

$$\rho = 3\mathrm{R}\frac{a-1}{a}\beta v + 3\mathrm{RT}_s \log_e a,$$

where for monatomic substances a is to be taken as 1.33; with $\beta v = 85$, calculation gives $\rho = 270$, in full agreement with Eucken's value.

For mercury, calculation gives 573 (instead of 555); the discrepancy shown by the latter substance is explained in Ratnowsky's table by the fact that βv has been taken as 61 instead of 96. The regularities discovered by Ratnowsky thus give us a further acceptable verification of these measurements.

The heat of evaporation of liquid argon may be deduced from Crommelin's figures by making use of D. Berthelot's formula,

$$pv = \mathrm{RT}\left\{1 + \frac{p}{\pi_0}\frac{9}{128}\tau(1 - 6\tau^2)\right\}$$

* "Verhandl. d. Deutsch. physik. Ges.," 1914, p. 1033.

by means of which equation **(81)** is transformed into

$$\lambda = 4\cdot 571 T^2 \frac{d \log p}{dT}\left\{1 + \frac{p}{\pi_0}\frac{9}{128}\tau(1 - 6\tau^2) - \frac{v'}{v}\right\};$$

v' for liquid argon is 0·03 lit., and v may be calculated from the gas laws since there is only a small correction concerned.

We thus have (for further details, cf. Nernst, 103) for the heat of evaporation of solid argon at the melting-point

$$\lambda = 1616(1 - 0\cdot 039) + 268 = 1821.$$

From the vapour pressure measurements on solid argon the value 1819 results for the same temperature, though there is some uncertainty as there is an error, large in Crommelin's first series and at any rate appreciable in his second (due to the admixture of neon or other volatile gas with the solid argon?).

We have therefore the following values for the heat of evaporation of solid argon at its melting-point (83·8°):—

Calorimetric measurement by Eucken .	1793 cals.
Vapour pressure curve of liquid argon by Crommelin with Eucken's heat of fusion	1821 ,,
Vapour-pressure curve of solid argon by Crommelin	1819 ,,

The following should be noted: Eucken's heat of evaporation at the boiling-point probably relates really, as he himself remarks (*loc. cit.*, p. 16), to a rather higher temperature, i.e. the first number is probably rather too small. The same error which affected Crommelin's measurements on solid argon at low temperatures must also have been operative in the same direction, though only to a slight extent, at higher temperatures, so that the last value in the above comparison would also be rendered too small: the middle number is doubtless the most reliable. It is therefore best not simply to take the mean; the most probable value, possibly somewhat too low, may be taken as 1817.

CALCULATION OF CHEMICAL CONSTANTS

For Eucken's atomic heat of solid argon the equation

$$C_p = f\left(\frac{85}{T}\right) + 0.00034 T^2$$

holds, where the Debye function is once more to be introduced as above. The energy content of solid argon at the melting-point is thus

$$E = \int_0^{83.8} C_p dT = 334 + 79 = 413,$$

whence we have

$$\lambda_0 = 1817 + 413 - 83.8 \times 4.963 + 11 = 1825;$$

the figure 11 represents, according to equation **(124)**, the change in internal energy of the saturated vapour.

We can now apply equation **(122)** to any temperature for which the vapour pressure p of solid argon is known. We select $T = 78.5$, since in this region Crommelin's measurements are apparently not yet appreciably affected by the source of error above mentioned, and, on the other hand, the gas laws hold here sufficiently well for the saturated vapour. For this temperature $\log_{10} p$ equals -0.496; hence, just as before,

$$-0.496 = -5.087 + 4.740 - 0.900 + C;$$
$$C = 0.75 \pm 0.06.$$

In the case of argon the uncertainty is doubtless considerably greater than with mercury; in particular, the heat of evaporation may be in error by some 1 per cent. Of the other terms only the figure 0.900 is uncertain by an amount which we may estimate as ± 0.02, so that as maximum error, i.e. one greater than is probable, we have ± 0.06. Further vapour pressure measurements at low temperatures are desirable on solid argon of the highest purity.

Hydrogen.—The specific heat of solid hydrogen is not known, but Kohner and Winternitz (98) succeeded in making quite a reliable calculation by an indirect method making use of the reaction

$$Hg + H_2O = HgO + H_2.$$

Since the calculations involved are somewhat complicated and the possibility of an arithmetical error was, of course, not excluded, Frl. Miething has, at my request, checked the purely numerical part of the work; as was to be expected, she could only confirm its correctness.

Frl. Langen, who has been engaged in other calculations of the same sort, did, however, come across a mistake on the theoretical side, though it is one which is of no consequence in practice. In the evaluation of the integral

$$\int_0^{273\cdot1} \frac{E}{T^2} dT$$

for mercury, the specific heat in the solid and liquid states was taken into account, but not the heat of fusion. At $T = 234\cdot4$ the heat of fusion becomes added to E, so that the value of $2\cdot065$ found for the above integral must be increased by

$$\frac{1}{4\cdot6} \int_{234\cdot4}^{273\cdot1} \frac{555}{T^2} dT = 0\cdot073,$$

i.e. to $2\cdot138$.

The final sum given by Kohner and Winternitz,

$$-34\cdot46 = -38\cdot21 + 6\cdot19 + 1\cdot92 - 1\cdot00 - 2\cdot06 + C,$$
$$C = -1\cdot30 \pm 0\cdot15,$$

leads us, after applying the above correction, to the slightly different final value,

$$C = -1\cdot23 \pm 0\cdot15.$$

In all other respects the authors appear to have adequately discussed the limits of error. Though a glance at the final sum shows that these limits may be fairly high, the authors rightly point out that the difference between the chief terms is fairly well known. Full reliance may therefore be placed on the final result.

Iodine.—A similar calculation was carried out by Stern *

* "Ann. d. Physik," (4), **44**, 497 (1914).

CALCULATION OF CHEMICAL CONSTANTS

for iodine vapour; the concentration of iodine atoms in the saturated vapour of solid iodine was determined from the dissociation, measured at high temperatures, and formula **(122)** was then applied. A fairly considerable deviation (0·83) was found from the theoretical value required by equation **(103)**. I was able to explain this, in part, by the fact that my measurements of the atomic heat of solid iodine at the lowest temperatures were appreciably affected by the transformation phenomenon mentioned on page 53; using Günther's figures, though these are provisional, the above difference is reduced to 0·57. A discussion of the remaining observations available shows that the difference eventually left is well within the error of observation (cf. here Nernst, 103). In any case, fresh measurements of the specific heat of rarefied iodine vapour, and also of the vapour pressure of solid iodine over a greater interval of temperature, would be eminently desirable for the further elucidation of the question; it would be simpler if the dissociation tension of a solid iodide could be measured over a wide range at high temperatures, where the vapour would be extensively dissociated.

Summary.—In the following table are collected the reliable results up to date * :—

	C	M	C_0
H_2	$-1·23 \pm 0·15$	2·016	$-1·69 \pm 0·15$
A	$0·75 \pm 0·06$	39·88	$-1·65 \pm 0·06$
Hg	$1·83 \pm 0·03$	200·6	$-1·62 \pm 0·03$

As we shall see in the following chapter, all theories advanced up till now agree in leading to the result that—

$$C = C_0 + 1·5 \log_{10} M,$$

where M is the molecular weight of the gas concerned. C_0

* See also Supplement, p. 266.

is calculated in the last column and, as may readily be seen, all the results may be satisfied by the value

$$C_0 = -1.62 \pm 0.03,$$

i.e. this number is within the present limits of error.

It may be recalled here once more that the limits of error indicated are not the probable error, but depend rather on a careful estimation of the degrees of uncertainty, and indicate what error is to be assumed as the maximum in the most unfavourable case. It results from this that any theory which leads to values of C_0 which lie appreciably beyond -1.59 to -1.65 is to be regarded as incomplete or wrong.

According to formula **(103)**,

$$C = \frac{i}{2 \cdot 303} = \log_{10} \frac{(2\pi m)^{3/2} k^{5/2}}{h^3}$$

or, introducing the numerical values (cf. Appendix and Nernst, 103, p. 194) and eliminating m by means of the equation

$$M = Nm$$

(M being the molecular weight),

$$C = C_0 + 1.5 \log M = -1.608 + 1.5 \log M \quad . \quad \textbf{(125)}$$

The above table, from which we have concluded the most probable value of C_0 to be -1.62 is thus a verification of the Sackur-Tetrode formula which we may well claim to be unexpectedly striking in view of the difficulty of the experimental test. For mercury in particular, for which by far the most accurate data are available, the agreement is perfect.

Although no longer surprising in view of the recent development of physics, the result of the investigations, both theoretical and experimental, which have been dealt with in this chapter appears to be a remarkable one; for the calculation of vapour-pressure curves and of chemical equilibria in which gases take part is rendered possible by means of Planck's constant h in addition to thermal data, i.e. by a quantity which was originally determined from radiation measurements.

The point which is most important in connection with the question under treatment in this book is that we are able to show that there is fresh strong evidence for the logical force of our Heat Theorem. This evidence lies in the experimental proof that there is really a general significance in the formula, deduced from a somewhat questionable mechanical model of a solid body and from molecular theory, for the calculation of constants of integration not fixed by thermodynamics.

4. Further Applications.—It is obvious that further equilibria, in particular those of a chemical nature, may now be calculated with the aid of the chemical constants which have been determined theoretically in what seems to be a perfectly reliable manner. In practice, however, we meet the difficulty that it is only in the case of hydrogen that the loss in rotational energy of a polyatomic molecule is at present known, so that all chemical processes are excluded for the time being other than the reaction,

$$H + H = H_2.$$

Isnardi (102a) has applied the Heat Theorem to this special case, making use of the theoretical chemical constants. Employing the value for the extent of dissociation measured by Langmuir, he has arrived at a figure of about 100,000 cals. for the heat of dissociation (reduced to absolute zero). A value very close to this was found independently by Langmuir [*] at the same time, in a recalculation of his measurements from the standpoint of the Second Law. This result is of great importance, for it cannot be reconciled with the requirement, made by the model of diatomic hydrogen proposed by Bohr and used with so much success by Debye, that this heat evolution [†] should amount to 60,000 cals. It

---

[*] " Zeitschr. f. Elektrochem.," **23**, 217, 1917.
[†] Cf. on this point Bohr, " Phil. Mag.," 1913, p. 863.

is probable, as suggested in my discussion (108), that there are other forms of energy at the absolute zero to be considered in the models of diatomic and monatomic hydrogen.

Had the heat evolution in the combination of two hydrogen atoms been as small as is required by Bohr's model, clear signs of dissociation must have been recognizable in the experiments of Pier and his followers (p. 17); this may be seen with certainty from the approximation formula **(92)** derived on page 137. The value 60,000 cals. is thus to be regarded as wrong.

The decrease in the energy of rotation is not at present known for other gases, and the only procedure possible is therefore to determine the value of the chemical constant for each gas from the calculation of some particular equilibrium in which the gas concerned takes part. We have a control in that we must find a positive value for $\beta\nu$ if we assume that the energy of rotation falls off, as with hydrogen, nearly according to Einstein's function (p. 58). For di- and poly-atomic gases the equation **(101)** takes the form

$$\log_e p = -\frac{\lambda_0}{RT} + 2 \cdot 5 \log_e T + \frac{1}{R}\int_0^T\frac{dT}{T^2}\int_0^T (C_p - 2 \cdot 5R)dT + i,$$

where i is given by formula **(103)** of page 166. Further measurements on the decrease of rotational energy at low temperatures would be of the greatest interest. The treatment of any particular gaseous equilibrium has, of course, to be carried out just as is explained on page 125, the expression

$$n\left\{2 \cdot 5 \log_e T + \frac{1}{R}\int_0^T\frac{dT}{T^2}\int_0^T (C_p - 2 \cdot 5\ R)dT + i\right\}$$

being introduced with the appropriate sign into equation **(73)** for each gaseous molecule which reacts with the number of molecules n.

The calculation of a number of examples by Frl. Langen,

CALCULATION OF CHEMICAL CONSTANTS

which is shortly to appear,* has led to the result that the loss of rotational energy must occur with NH_3 and H_2O at temperatures readily attainable, but that this does not seem to be the case for CO, N_2, O_2, or NO.

The chemical constant of the negative electron is calculable from formula **(103)**; the thermal dissociation of an atom into the positive ion and the negative electron is thus definitely determined if we know the heat of dissociation. Now, in certain cases this is given by Bohr's model of the atom, though this method is, it is true, somewhat hypothetical at the present time.

* [Langen, "Zeits. f. Elecktrochem.," **25**, 25 (1919).—TR.]

CHAPTER XIV

DIRECT APPLICATION OF THE HEAT THEOREM TO GASES

1. Statement of the Problem.—In order to get a clear idea of what remains to be dealt with in this chapter, we must briefly review the results so far obtained.

In Chapter X we applied our Heat Theorem to equilibria in which gases take part, but this application was only indirect, in that we started from condensed systems and operated thereafter solely with classical thermodynamics. We obtained the result that any equilibria, even in gaseous systems, are calculable from thermal data if there is available, for each molecular species which does not also take part as a condensed phase, a measurement of any other equilibrium, in the simplest case a measurement of vapour pressure.

By introducing the conception of chemical constants a simple and clear formulation was given to this result.

We have seen in the last chapter that the chemical constant may be calculated according to the formula

$$C = -1·608 + 1·5 \log M \quad . \quad . \quad (125)$$

for monatomic gases, and also for polyatomic gases when they have been brought by cooling into the "thermally monatomic" state; the molecular weight M is, of course, to be found from a simple density determination.

It was assumed, however, that all gases behave like hydrogen as regards their molecular heat. A behaviour such as hydrogen actually exhibits was foreseen (Chapter V) from theory and was also found in the first example

DIRECT APPLICATION TO GASES

investigated ; the general occurrence of this behaviour for all gases can therefore hardly be doubted.

The question of the theoretical calculation of gaseous equilibria is thus solved in principle, subject, of course, to the further assumption that formula **(125)** is also of general validity ; but this, again, we can hardly doubt in the light of our present knowledge.

The calculation which we are discussing requires, therefore, for its performance a knowledge of

1. All the thermal data which come into consideration (heats of combination, specific heats) ;
2. The vapour densities of the respective gases, for the determination of M.

Although the molecular weight will be known in practically all cases, we recognize in the introduction of the second condition an indication that the consequences of our Heat Theorem have not yet been fully explored.

As a matter of fact, we have, even in the last chapter, always applied the Theorem to condensed systems only and not to the gas itself ; we may readily convince ourselves that, in the method of treatment hitherto adopted, it does not hold, strictly speaking, for gaseous systems.

Consider the expansion of a gas from the volume v_1 to the volume v_2 ; then, if the gas laws hold without restriction for all temperatures,

$$A = RT \log_e \frac{v_2}{v_1}, \quad \lim \frac{dA}{dT} = R \log_e \frac{v_2}{v_1} \text{ (for } T = 0\text{)} \quad (126)$$

i.e. $\frac{dA}{dT}$ for this process does not converge towards zero in the neighbourhood of the absolute zero as our Heat Theorem requires.

Consider an equilibrium in which gases take part, e.g. one in a homogeneous gaseous system ; then (cf. p. 188)

$$A = -RT \log_e K = U_0 - \Sigma nT \times 2 \cdot 5 R \log_e T$$
$$- \Sigma nT \int_0^T \frac{dT}{T^2} \int_0^T (C_p - 2 \cdot 5 R) dT - RT \Sigma ni \quad . \quad (127)$$

and at low temperatures

$$\frac{dA}{dT} = -\Sigma n \cdot 2\cdot 5 R \log_e T - \Sigma n \cdot 2\cdot 5 R - \Sigma n i R \quad . \quad (128)$$

In equation **(127)**, U_0 denotes the heat evolution of the reaction at the absolute zero, and E the energy content of each separate gas ; the summation is to be taken as described on page 126. The expression in formula **(128)** is, however, by no means zero for $T = 0$, i.e. our Heat Theorem in this case again does not hold, in its direct application to gases. This is, of course, not surprising because we have so far assumed the applicability of the gas laws down to the lowest temperatures, which, according to equation **(126)** precludes the applicability of our Heat Theorem.

So far as concerns the more important practical applications of the Theorem, these considerations are of little moment ; equation **(127)** does, in truth, contain the complete solution of the question how to calculate gaseous equilibria thermodynamically (cf. also p. 187).

Matters are different, however, when we are inquiring into the general validity of the Heat Theorem ; it is then unsatisfactory from the point of view of principle (not of practice) if it cannot be applied directly to gases. The question narrows itself down to this :—

Does equation **(126)** continue to hold at the lowest temperatures ?

We have already seen in Chapter V that this cannot be the case if we regard the requirements of the quantum theory as generally valid ; we have seen, too, that there are already direct observations on helium (p. 71) which indicate an abnormal behaviour of the gas at very low temperatures, and here, apparently, is the key to a closer understanding. We shall now deal more fully with the theory of the " degeneration of gases " with which many eminent theorists, as Tetrode, Sackur, Keesom, Sommerfeld, Planck, have occupied themselves of late years, agreeing among themselves in principle though differing in details. I give the description which I myself have developed, for, as far as I can see, it is the only one of which the consequences have never collided with the facts.

2. General Theory of the Degeneration of Gases. Fundamental Assumptions (cf. Nernst, 95a).—The following considerations are to be regarded as a sort of prolegomena to any further special theory of the "Degeneration of Gases," for they rest, I think, everywhere on a sure basis.

We make, first, the two following fundamental assumptions only :—

1. Every gas, when cooled at constant volume with exclusion of condensation, eventually reaches a state in which the heat capacity is negligibly small.
2. The decrease in specific heat of a given mass of gas occurs the earlier, the greater its density is.

Fig. 20 indicates the behaviour of a mol of any gas of volume v which is either monatomic or has been deprived of its rotational energy by cooling; we have

$$v_1 > v_2$$

The above two fundamental assumptions are contained in the work of all theorists who have hitherto occupied themselves with this question; from the experimental side, too, they have received weighty support in Eucken's discovery (p. 71) that the molecular heat of helium begins to fall more quickly below the value $\frac{3}{2}R$, the greater its density. Reference may also be made to the interesting results of Polányi,* who also arrived, from theoretical considerations, at the conclusion that all bodies must, when subjected to sufficiently high pressure, reach a state in which their energy content is negligibly small; applied to

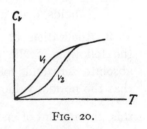

FIG. 20.

---

* " Verh. Deutsch. phys. Ges.," **15**, 157 (1913).

gases above the critical point, this means that under extremely high compression they must behave thermodynamically like solid bodies at the absolute zero.

In these circumstances, the above assumptions may be regarded, if not as reliable experimental facts, at any rate as very probable working hypotheses.

Now, if the heat capacity vanishes at low temperatures, the application of the Heat Theorem must also be regarded as safe; hence the considerations previously developed (Chapter VII) concerning the impossibility of attaining the absolute zero are to be transferred to the present case. We thus arrive of necessity at the third fundamental assumption:

3. For every physical or chemical change which is associated with a maximum expenditure of work A, the law

$$\lim \frac{dA}{dT} = 0 \text{ (for } T = 0) \qquad . \qquad . \quad \textbf{(129)}$$

holds, even in the case of gases of finite density

3. Application of Classical Thermodynamics.—Consider the following cyclic process: we warm a gas from the absolute zero and volume v_1 to a temperature T' so high that the normal value of specific heat $\frac{3}{2}R$ is reached; to do this, the amount of energy $\int_0^{T'} C_{v_1} dT$ must be supplied. At the temperature T', the gas being now in the ideal state, we bring it to the volume v_2; there is, of course, no change in energy associated with this. The gas is then cooled at the volume v_2 to the absolute zero, which results in the energy, $\int_0^{T'} C_{v_2} dT$ being given out. Finally, the gas is brought at the absolute zero from the volume v_2 to the volume v_1, the heat evolution associated with this being $U_1^0 - U_2^0$. The First Law gives us

$$-\int_0^{T'} C_{v_1} dT + \int_0^{T'} C_{v_2} dT + U_1^0 - U_2^0 = 0.$$

DIRECT APPLICATION TO GASES

According to the Second Law, the changes of the total and of the free energy are the same at the absolute zero; we may therefore put

$$A_2^0 - A_1^0 = U_2^0 - U_1^0 \quad (T = 0),$$

or

$$A_2^0 - A_1^0 = \int_0^{T'} (C_{v_2} - C_{v_1}) dT \quad . \quad . \quad (130)$$

Since at all temperatures $C_{v_1} > C_{v_2}$, we arrive at the remarkable result that, even at the lowest temperatures, a gas must do work in expanding. If we denote by p the pressure of the gas at the absolute zero and at volume v we have, of course, also

$$A_2^0 - A_1^0 = \int_{v_1}^{v_2} p\, dv \quad . \quad . \quad . \quad (131)$$

4. Application of the New Heat Theorem.—The new Heat Theorem gives

$$\lim \left(\frac{dp}{dT}\right)_v = 0 \quad (\text{for } T = 0) \quad . \quad . \quad (132)$$

It follows also from this that gases at very low temperatures and at constant pressure have a volume independent of the temperature.

The exceptional position which we have hitherto had to assign to gases as regards the application of the Heat Theorem has now disappeared; we may now deal with gases just as we have done for condensed systems in Chapter IX. There is, it is true, the disturbing difference that, though we know the variation of molecular heat C_v at very low temperatures for many solids, we do not know it at present for a single gas. But perhaps the special theory of the "degeneration of gases" which is to be discussed below may fill this gap.

In any case, it will be of interest to inform ourselves how the calculation for gases is to be carried out; we select as the most important example of the kind the vapour-pressure curve.

For the sake of simplification we again consider temperatures so low that we may regard the specific heat of the condensate as negligibly small. This does not entail any limitation, for the specific heat may be readily taken into account if necessary. We assume, further, that the volume of the saturated vapour is large compared with that of the condensate, an assumption which is of course perfectly justified at low temperatures. Then, according to the Second Law, we have for the vapour pressure the formula which has already been often employed,

$$\log_e p = -\frac{\lambda_0}{RT} + 2 \cdot 5 \log_e T + i \quad . \quad . \quad (101)$$

Further, at the saturation point, we have for the maximum work A′ associated with the condensation

$$A' = -pv = -RT' \quad . \quad . \quad (133)$$

We now apply the new Heat Theorem to the formation of one mol of the solid. If we denote by λ_0 the heat evolved in the condensation of a mol of a very rarefied gas, referred to the absolute zero, the corresponding quantity for the volume v (cf. equation **130**) amounts to

$$\lambda_{0,v} = \lambda_0 - \int_0^{T'} (C_v - \tfrac{3}{2}R) dT,$$

and the variation with temperature is given by

$$\lambda_v = \lambda_{0,v} + \int_0^T C_v dT.$$

The variation of A is hence also definitely determined, and for each v not only is A′ fixed for the equilibrium point, but also A for every temperature. We thus come at once to a much more general treatment of the problem than is given

DIRECT APPLICATION TO GASES

by formula **(101)**, quite independently of whether the integration constant i disappears * just as it does for condensed systems.

Our Heat Theorem gives at once for A

$$A = - T\int^{T}\frac{\lambda_v}{T^2}dT\,;$$

the temperature T' at which solid and gas are in equilibrium with one another is given by the condition

$$A = A',$$

or, making use of **(133)**,

$$\int^{T'}\frac{\lambda_v}{T^2}dT = R \qquad . \qquad . \qquad . \qquad \textbf{(134)}$$

The behaviour is again most clearly shown by a graphical representation; the λ-curves are drawn in Fig. 21 for two volumes, assuming once more $v_1 > v_2$. From the shape of the two λ-curves the corresponding A-curves are also definitely fixed by the Heat Theorem; their intersection with the A'-curve (broken line) indicates the temperature at which vapour of volume v is just saturated. As this point of intersection falls in the region where the ordinary gas laws hold (cf. *infra*), the vapour pressure p may be calculated from v by

$$pv = RT.$$

We thus see that the vapour-pressure curve may actually be derived solely from thermal data.

By way of further explanation, the following comments may be made :—

1. The diagram of Fig. 21 corresponds perfectly with that of the fusion process (Fig. 13, p. 102); the U-curve bends upwards, while the A-curve bends downwards; equilibrium occurs in general at

$$- A = p(v - v')\,;$$

* Cf. also Tetrode, " Physik. Zeitschr.," **14**, 212 (1913).

whereas, however, A' at the melting-point becomes in practice vanishingly small (p. 102), this is no longer the case for the equilibrium between vapour and condensate. Moreover, strictly speaking, $\lambda_{0,v}$ is somewhat smaller for the greater volume v_1 than for the smaller v_2; this difference, which is of practically no importance, is not expressed in Fig. 21.

2. As above assumed, the gas laws continue to hold for the saturated vapour even at the lowest temperatures; this is because, as the temperature decreases, the vapour becomes

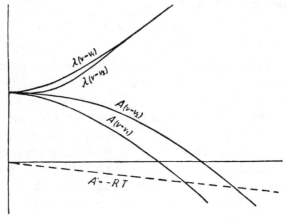

Fig. 21.

extraordinarily rarefied, and thus permanently preserves itself from "degeneration." We shall deal again with this important behaviour when we discuss the special theory.

3. The fact that the integration constant i has disappeared is to be interpreted as meaning that though formula **(101)** continues, according to comment 2, to be perfectly exact down to the lowest temperatures, there is an additive constant which appears in making the integrations,

$$E = \int_0^T C_v dT, \quad A = \lambda_{0,v} - T\int_0^T \frac{E}{T^2} dT,$$

for higher temperatures ; this is always before the intersection with the A' line. This additive constant must, of course, be identical with i, and must lead to formula **(103)**.

The latter condition indicates a definite line for any special theory of the degeneration of gases.

5. The Physical Behaviour of Gases at Low Temperatures.—According to the above considerations, it must be possible to convert any gas, if strongly cooled at constant volume, with exclusion of condensation, into a state in which it has a great similarity to the amorphous state ; high compression at constant temperature has an effect similar to that of intense cooling.

This conception finds, perhaps, some experimental support in the results of seismological investigation of late years ; the conclusion has here been reached that transverse vibration can be propagated through the interior of the earth. Now, in view of its enormous temperature, the interior of the earth will contain only gases, chiefly iron vapour, above their critical temperature, so that gases must be credited with the capability of assuming, under intensely high compression, certain properties of solids.

Let us once more picture to ourselves how a rarefied gas will behave, on the above reasoning, when we compress it at a constant, extremely low temperature. We shall imagine that with this compression the gas passes continuously into the liquid state, which is here, of course, identical with the amorphous or glassy state. At first work must be supplied for the compression, just as though there were forces of repulsion between the molecules. But later, when the compression is high enough, very strong forces of attraction must be manifested which occasion the heat of condensation, very large compared with the above work of compression. Finally, when the body has been converted into the ordinary amorphous state, it resists further compression, i.e. there appear again very strong forces of repulsion. It must, of course, remain a matter for enquiry what is the nature of the manifestations of force which we have mentioned, and whether they are in fact what would be called central forces in physics ; I should, therefore, have introduced these forces only as a mnemonic aid ; but it is certain that ordinary thermal motion plays no part here.

6. A Special Theory of the Degeneration of Gases.*—

Such a theory is only to be obtained from considerations of molecular theory, and by going into them we shall be leaving our particular subject. But, on the other hand, the question dealt with in this chapter is of such great and fundamental importance that any further support or extension of our conception of the " degeneration of gases " is of the greatest interest, even from the purely thermodynamical standpoint.

Theories of this kind have been developed by Tetrode,† Sommerfeld,‡ and Keesom,§ all on the assumption that a gas behaves at low temperatures like a solid and that therefore Debye's well-known theory of specific heats becomes applicable in its essentials.‖

Tetrode, who led the way in an extremely original manner, expressly remarks, however, that the application of Debye's method to gases and liquids is to be considered much less safe than it is with solids, and he is, therefore, inclined to ascribe a rather qualitative character to such considerations.

Personally, I have never been able to recognize any particular physical nucleus in this method of treatment ; where are the forces to come from at low temperatures to

---

* See also Supplement, p. 267.

† " Physik. Zeitschr.," **14,** 212 (1913).

‡ " Göttinger Vorträge," p. 134 (Teubner, Leipzig, 1914).

§ " Physik. Zeitschr.," **15,** 695 (1914).

‖ Planck, in an interesting attempt to develop the theory of the " degeneration of gases " on the assumption that the impacts of atoms take place according to the laws of ordinary mechanics, arrives at apparently impossible results (cf. " Sitzungsber. d. Preuss. Akad. d. Wiss.," 1916, p. 665).

DIRECT APPLICATION TO GASES

confer the properties of solids on even extremely rarefied gases?

We may readily prove, moreover, that this conception does not satisfy the facts quantitatively; for Tetrode finds $C_0 = -2\cdot35$, Sommerfeld $C_0 = -3\cdot25$, and Keesom, who has been concerned to reduce the existing discrepancies by introducing the "energy at the absolute zero" (Nullpunkt energie), finds $C_0 = -2\cdot4$. But all these values are irreconcilable with the table on page 185.

I have recently (108) developed an essentially different conception of what the degeneration of gases is; according to this the translatory motion of the molecules of gas is converted more and more at low temperatures, under the influence of the absolute zero radiation (Nullpunkts-strahlung) of the ether, into a circular movement, such that at the zero itself the molecules of gas revolve with constant velocity about equilibrium positions, uniformly distributed in space.

I expressly designated the above special theory of degeneration, when it appeared, as provisional; but since it seems probable that it comes at any rate near the truth, I may here give briefly the formulæ to which it leads, and extend it in a few respects.

In the volume V let there be N molecules ($= 1$ mol); the volume available for each separate molecule then amounts to

$$v = \frac{V}{N}; \qquad \qquad (135)$$

for the radius r of the circle in which it rotates with frequency ν, I found by a simple treatment (not conclusive, it is true)

$$r^2 = \frac{3}{\pi} v^{2/3} \qquad \qquad (136)$$

and for the frequency of rotation

$$\nu = \frac{h}{4\pi m v^{2/3}} \qquad \qquad (137)$$

From the assumptions set out in the above-mentioned paper, the energy content of the gas comes out as

$$E = \frac{3}{2}N \frac{\frac{h^2}{4\pi m v^{2/3}}}{e^{4\pi m v^{2/3} kT} - 1} \qquad . \quad . \quad (138)$$

The reliability of this equation depends, of course, on whether we have made a correct guess in our model of a degenerated gas. But a different model would only lead to a different, and not very essentially different, factor in place of 4π.

Any such equation contains, it is true, the full theory of the gaseous state, as Tetrode (*loc. cit.*) and Keesom (*loc. cit.*) have shown; our equation is distinguished only by its simplicity. A direct experimental test of equation **(138)** will hardly be possible, because for rarefied gases, to which alone it refers, the decrease of specific heat below the normal value takes place at much too low temperatures. But indirectly there is an important application.

For we can now write down, without neglecting anything, the vapour pressure of any system which gives off a monatomic gas. Let λ_0 be the amount of heat which is developed when the infinitely rarefied gas condenses at the absolute zero; if we consider its condensation from the finite volume V, we then have

$$U_0 = \lambda_0 + \frac{3}{2}Nh\nu = \lambda_0 + \frac{3}{2}N\frac{h^2}{4\pi m v^{2/3}} \qquad . \quad (139)$$

At the temperature T this quantity becomes

$$U = \lambda_0 + \frac{3}{2}N\frac{h^2}{4\pi m v^{2/3}} + E \qquad . \quad . \quad (140)$$

where E denotes the energy content of the gas (at the constant volume V). In order not to make our formulæ unnecessarily complicated, we again assume the energy content of the condensate to be negligibly small.

DIRECT APPLICATION TO GASES

Application of the new Heat Theorem results at once in the two equations

$$U = U_0 + E \qquad \qquad (141)$$

$$A = U_0 - T\int_0^T \frac{E}{T^2}dT \qquad \qquad (142)$$

Since U_0 (equation **139**) and E (equation **138**) are known, the two sets of curves U (or λ) and A of Fig. 21 (p. 198) are completely determined.

The equilibrium of the saturated vapour is given by the equation

$$A' = -pV \qquad \qquad (143)$$

Since p can be definitely fixed by V and T from equation **(143)**, the vapour-pressure curve is defined by **(142)** from the absolute zero upwards. The last term of equation **(142)** amounts, as may easily be found with the aid of **(138)**, to

$$\tfrac{3}{2}RT \log_e \left(e^{\frac{h^2}{4\pi m v^{2/3}kT}} - 1\right) - \tfrac{3}{2}N\frac{h^2}{4\pi m v^{2/3}}.$$

Now, as a matter of experience, the gas laws hold for saturated vapours at sufficiently low pressures, i.e. at sufficiently low temperatures; we may therefore put

$$A' = -RT.$$

Thus, after a few simple transformations, and writing also for the constant volume

$$v = \frac{V}{N} = \frac{RT}{pN},$$

we obtain finally from equation **(142)**

$$\log_e p = -\frac{\lambda_0 + \tfrac{3}{4}N\dfrac{h^2}{4\pi m v^{2/3}}}{RT} + \tfrac{5}{2}\log_e T + \log_e \frac{(4\pi m)^{3/2}k^{5/2}}{eh^3} \qquad (144)$$

In the above equation e is the base of the natural logarithms.

Comparing this with the vapour pressure formula **(101)**, page 166,

$$\log_e p = -\frac{\lambda_0}{RT} + 2\cdot 5 \log_e T + i, \qquad \textbf{(101)}$$

we recognize that the two formulæ are identical except that in **(144)** the expression $\lambda_0 + \tfrac{3}{4}N\dfrac{h^2}{4\pi m v^{2/3}}$ occurs, instead of λ_0. The second part of the former expression is, however, only a very small correction term, and vanishes entirely at large values of $v = \dfrac{V}{N}$, i.e. at low temperatures, so that equations **(144)** and **(101)** are actually identical for low temperatures.

It will be useful to convince ourselves of the correctness of what has just been said by working out a numerical example, and such a calculation will also give us an idea of the deviations from the gas laws which result from degeneration.

Consider a gas of atomic weight 1, so that

$$\nu = \frac{hN^{2/3}}{4\pi m V^{2/3}} = 0\cdot 2915 \times 10^{10}, \quad \beta\nu = 0\cdot 1417°;$$

where what is known as the normal volume of a mol, i.e. 22,412 c.c., has been put for V. Under the same conditions the second term on the right of equation **(140)** amounts to

$$\tfrac{3}{2}Nh\nu = 0\cdot 422 \text{ gm. cals.}$$

For hydrogen (atomic weight equals 2) the expression concerned thus amounts to

$$\tfrac{3}{4}N\frac{h^2}{4\pi m v^{2/3}} = \tfrac{3}{4}Nh\nu = 0\cdot 105 \text{ gm. cals.,}$$

a value which is very small compared with λ_0 (ca. 250 cal.). Moreover, it relates to a vapour pressure of ca. $1/14$ atm., and at smaller pressures, i.e. lower temperatures, it very rapidly becomes quite negligibly small

DIRECT APPLICATION TO GASES

We find for the integration constant i, by a method quite different from that of the previous chapter (p. 173),

$$\iota = \log_e \frac{(4\pi m)^{3/2} k^{5/2}}{eh^3} \text{ instead of } \log_e \frac{(2\pi m)^{3/2} k^{5/2}}{h^3}. \quad (145)$$

The two expressions are nearly identical, differing only a little in the numerical factor; for

$$\log_e \frac{(4\pi)^{3/2}}{e} = 1\cdot 215 \text{ and } (2\pi)^{3/2} = 1\cdot 197,$$

a difference which is negligible with the present accuracy of measurement (cf. p. 186).

Probably the factor found by Sackur, Tetrode, and Stern will be the more correct; nevertheless, our derivation of the constant of integration i, even though only approximately true, constitutes a step forward, in that it gives us the shape of the complete curves for U and A for the condensation of a mass of gas; as equation **(142)** shows, it brings to our mind the fact that i results from the variation of the energy content or specific heat of the gas.

7. Equation of State for Ideal Gases at Very Low Temperatures.—If, as we have done throughout, we denote the negative energy content of the system by U, the formulæ of the previous section show that at the absolute zero

$$-U_0 = \tfrac{3}{2} N h \nu,*$$

and more generally

$$-U = \tfrac{3}{2} N h \nu + E \quad . \quad . \quad . \quad (146)$$

where ν is determined by formula **(137)** and E by formula **(138)**. Our Heat Theorem gives

$$-A = \tfrac{3}{2} N h \nu - T \int_0^T \frac{E}{T^2} dT, \quad . \quad . \quad (146a)$$

and, making use of equation **(53)**, page 95, it follows at once that

$$-A = \tfrac{3}{2} RT \log_e \left(e^{\frac{\beta \nu}{T}} - 1 \right) * \quad . \quad . \quad (146b)$$

THE NEW HEAT THEOREM

Now, at constant temperature,
$$pdV = dA,$$
or
$$p = \frac{\partial A}{\partial V}. \quad . \quad . \quad . \quad (147)$$

Therefore, if we partially differentiate equation **(146b)** with respect to V, it follows that

$$p = -\tfrac{3}{2}R \, \frac{\beta e^{\frac{\beta \nu}{T}}}{e^{\frac{\beta \nu}{T}} - 1} \frac{\partial \nu}{\partial V}$$

and making use of the equation

$$\nu = \frac{hN^{3/2}}{4\pi m V^{3/2}}, \quad . \quad . \quad . \quad (147a)$$

we obtain finally, after a few simple transformations,

$$p = \frac{R}{V}\frac{\beta \nu}{1 - e^{-\frac{\beta \nu}{T}}} \overset{*}{} \quad . \quad . \quad . \quad (148)$$

If our theory is strictly correct, equation **(148)** is obviously of a fundamental character, for it has to apply in place of the equation of state hitherto generally assumed for ideal gases,

$$pV = RT; \quad . \quad . \quad . \quad (149)$$

as a matter of fact, equation **(148)** reduces to **(149)** both for high values of T and for large values of V.

But even if our formulæ describe the phenomenon of the degeneration of gases to a first approximation only, equation **(148)** may serve, at any rate, for preliminary exploration.

For regions of temperature in which T is large compared with $\beta \nu$ we have

$$pV = \frac{RT}{1 - \dfrac{\beta \nu}{1 \cdot 2 \cdot T} + \dfrac{(\beta \nu)^2}{1 \cdot 2 \cdot 3 \cdot T^2} - \cdots},$$

---

* These equations are incorrect. See Supplement, p. 268.

or, simplifying still further,

$$pV = RT\left(1 + \frac{\beta v}{2T}\right) \quad . \quad . \quad . \quad (150)$$

For a gas of atomic weight 1 at the normal volume of a mol we found above

$$\beta v = 0.1417° ;$$

thus for hydrogen, which is under a pressure of 1 atmosphere at the temperature of melting ice, the equation of state for any temperature at constant volume is

$$pV = RT\left(1 + \frac{0.0354}{T}\right) \quad . \quad . \quad (151)$$

If we are dealing, not with the normal volume of a mol (22,412 c.c.), but with the volume V, the correction term in equation **(151)** has to be multiplied by $\left(\frac{22410}{V}\right)^{2/3}$. If, finally, we introduce p instead of V from **(149)**, which is allowable so long as we are dealing with a small correction term, equation **(151)** takes the form

$$pV = RT\left(1 + \frac{A}{M}\frac{p^{2/3}}{T^{5/3}}\right) \quad . \quad . \quad (152)$$

where A is independent of the nature of the gas and is determined by equations **(147a)** and **(150)**.

For gases under low compression it is known that Daniel Berthelot's[*] equation of state, which is a modification of van der Waal's equation, holds to a good approximation:—

$$pV = RT\left\{1 + \pi\frac{9}{128}\tau(1 - 6\tau^2)\right\}; \quad \tau = \frac{\theta_0}{T}, \quad \pi = \frac{p}{\pi_0} \quad (153)$$

(θ_0 = critical temperature, and π_0 = critical pressure). This equation holds, as Berthelot showed, even for hydrogen, at any rate well above the critical temperature. Now as regards both the influence of pressure and the influence of

---

[*] Cf. "Theor. Chem.," pp. 256 *et seq.*

temperature, formulæ **(152)** and **(153)** are of very different construction, and, moreover, they take account of influences of very different nature ; this is obvious in that the corrections to be applied to the gas laws in virtue of the phenomena of degeneration are inversely proportional to the molecular weight, whereas the van der Waals' effects show, of course, an obvious tendency to become more pronounced as the molecular weight of the gas increases.

Since equation **(153)** holds well (at any rate considerably above the critical temperature) we must obviously conclude that the correction depending on degeneration is small in comparison with the correction determined by equation **(153)**. Any theory in which this condition is not fulfilled must therefore be rejected.

Of all the gases at present available to us, hydrogen must show the phenomena of degeneration most clearly, and a failure of formula **(153)**, which is purely empirical, would be expected to occur soonest here. As above remarked, it does not fail, or, at any rate, not definitely,* so that we may conclude with some degree of probability that any theory of degeneration is at fault which leads to correction terms commensurate with the observations at temperatures much above the critical temperature.

The deviation of hydrogen under normal conditions from the state of an ideal gas amounts to 0·00060 (Berthelot, *loc. cit.*, p. 40), which agrees well enough with equation **(153)**. Sackur's theory of degeneration gives 0·0005, much too high in the light of the above reasoning ; Keesom's theory gives the value 0·00024, which is still rather open to criticism ; our formula **(151)** estimates the deviation as 0·00013, actually considerably less than the observed number.

Once more, as on page 201, we have the result that, of the different theories which are quite analogous in the con-

* Cf. D. Berthelot, " Sur les thermomètres à gaz.," Paris, 1903. Gauthier-Villars.

struction of their formulæ, only that here developed accords with the facts without being forced.

For small deviations it should be justifiable to assume that equations **(152)** and **(153)** may be superposed, though it must be noted that the critical data of gases of low molecular weight must be considerably modified by degeneration. On this view the deviation from the Boyle-Mariotte law would give a curve concave towards the axis of pressure. Actually, Sackur * has been able to detect such a curvature for hydrogen and helium. More reliable evidence of degeneration should be obtainable from a detailed comparison of the three monatomic or quasi-monatomic gases, hydrogen, helium, and neon at low temperatures.

In conclusion, we may now answer the question already discussed on page 206, as to whether the gas laws hold for the saturated vapour, even at the lowest temperatures. We know that this is the case at low temperatures; according to equation **(150)**, the correction term to be applied owing to degeneration is inversely proportional to $TV^{2/3}$, and it may be seen, from any vapour-pressure curve or from a calculation on the lines of equation **(101)**, that this product increases very rapidly as the temperature falls. The molar volume of saturated vapour increases very much more rapidly than the absolute temperature decreases, so that as the temperature is lowered the saturated vapour becomes further and further removed from the influence of degeneration. If the light gases (nebulium, etc.) supposed by some astronomers to exist, were available to us, it would of course be very much easier to investigate the degeneration of gases experimentally; even monatomic hydrogen would offer great advantages over other gases.

Just as we arrived at the equation of state **(148)** with the aid of the hypothesis **(138)**, we should be able to obtain a generalized equation of state for gases, in other words, a

* " Zeitschr. f. Elektrochemie," **20**, 563 (1914).

generalized and perfectly exact van der Waals' equation, if we knew accurately the energy content E or, which comes to the same thing, the specific heats as dependent on the temperature and density of the gas. By a generalized equation of state we mean one which embraces the alterations in the classical equation **(149)** conditioned not only by degeneration, but by all the other influences. It is perhaps of interest to remark that the simple behaviour which I ascribed * to ideal concentrated solutions more than twenty years ago must occur for all gaseous mixtures at sufficiently low temperatures ; this theory, which I may call the forerunner of my Heat Theorem, has thus a wide field of application at low temperatures for any and every kind of mixture.

* " Wied. Ann.," **53**, 57 (1894).

CHAPTER XV

GENERALIZED TREATMENT OF THE THERMODYNAMICS OF CONDENSED SYSTEMS

IN this chapter we shall develop a rather more general conception of the application of the older laws of thermodynamics and of the new Heat Theorem. We shall limit ourselves essentially to condensed systems; we have dealt with gases sufficiently fully in Chapters X and XIII. Moreover, for anyone who accepts the theory of degeneration described in the previous chapter there is no longer any difference, as far as the application of our Heat Theorem is concerned, between condensed and gaseous systems. The substance of the following discussion may be found in a paper (93) laid before the Prussian Academy of Sciences in 1913.

1. Formulation of the Second Law for Condensed Systems.—We again write for the maximum work A isothermally obtainable, and the heat evolution U associated with it,

$$A - U = T\frac{dA}{dT}; \quad . \quad . \quad . \quad (1)$$

the differential quotient $\frac{dA}{dT}$ is to be determined in such a way that no performance of work is associated with the warming of the system.*

If we consider, for example, a body expanding in volume by dv, we have

$$A = p\,dv, \quad U - A = dQ, \quad \frac{dA}{dT} = \left(\frac{\partial p}{\partial T}\right)_v dv$$

---

* Helmholtz, " Ges. Abh.," II, p. 978.

and the heat developed in the expansion is

$$dQ = -T\left(\frac{\partial p}{\partial T}\right)_v dv \quad . \quad . \quad . \quad (154)$$

In this simple case the differential quotient $\frac{dA}{dT}$ is found if we imagine that the body in question is warmed up by dT at constant volume, i.e. without the performance of external work, and that the increase of pressure so caused is measured. It may be definitely pointed out here that, as is usual in thermochemistry, we reckon the heat given off from the system and therefore, of course, the external work supplied by the system, as positive, contrary to the sign frequently employed.

Now let the course of the process considered be determined by the special parameters w_1, w_2, ..., as well as by the volume v, so that we may write

$$A = K_1 dw_1 + K_2 dw_2 + \ldots + p\,dv \quad . \quad (155)$$

$$U = k_1 dw_1 + k_2 dw_2 + \ldots + p\,dv - T\frac{\partial p}{\partial T}dv \quad (156)$$

The establishment of the nature of the factors of proportionality is a problem for the detailed physical or chemical investigation of the system concerned. Then it follows from equation **(1)** that since the values of dw are to be made small, but may be selected as we please,

$$K_n - k_n = T\frac{\partial K_n}{\partial T} \quad . \quad . \quad . \quad (157)$$

Both the differential quotient $\frac{\partial p}{\partial T}$ and $\frac{\partial K_n}{\partial T}$ are to be taken * at constant volume and at a constant value of w.

Equation **(157)** is identical with that (1*d*) in Helmholtz, *loc. cit.* When applied to galvanic cells, for example, K_n

---

* The necessity for this condition will best be realized by considering the cyclic process involved; cf. " Theor. Chem.," pp. 25 *et seq.*

GENERALIZED TREATMENT

is equal to the E.M.F., $k_n d\epsilon$ to the change in energy observed when the quantity $d\epsilon$ of electricity is supplied by the cell at constant volume, and $\frac{\partial K_n}{\partial T}$ is equal to the temperature coefficient of the cell measured at constant volume.

In the practical application of equation **(157)** in the above important case there arises, however, a difficulty in that the temperature coefficient of the E.M.F. of galvanic cells has always been measured at constant pressure; the determination at constant volume is hardly practicable directly.

The following reasoning shows also how inconvenient is the introduction of the temperature coefficient at constant volume: if we warm a cell at constant volume, the pressure increase, and thus also the E.M.F., is definitely fixed only if we know, in addition to the coefficients of thermal expansion of the cell components, the proportions also in which they are present; the E.M.F. measured at constant pressure is, however, independent of the proportions of the separate phases. This fact again shows that equation **(157)** is of a form which cannot usually be employed in practice.

On a point of history it may be remarked, in this connection, that neither Helmholtz himself nor his pupil Czapski, nor later workers (Jahn and others), in applying equation **(157)** to galvanic cells, dealt with the change of E.M.F. at constant volume; they put, rather,

$$E - W_p = T\left(\frac{dE}{dT}\right)_p \qquad . \qquad . \qquad . \qquad \textbf{(158)}$$

The E.M.F. was practically always measured at atmospheric pressure; for W_p, the heat evolution at constant pressure, were introduced the data, e.g. of Thomsen, which also relate to the heat developed in a calorimeter when the reaction takes place under atmospheric pressure. I have been unable to find an explanation of this either by Helmholtz, Czapski, or their immediate followers; probably it was tacitly assumed (as it happens, rightly) that, in galvanic cells containing no gaseous phase, $\left(\frac{dE}{dT}\right)_p$ and $\left(\frac{dE}{dT}\right)_v$ may be regarded as practically equal, k_n of formula **(157)** and W_p being also only extremely slightly different.

A thoroughly exhaustive treatment of the question is given by R. Lorenz and M. Katayama,* who demonstrate the following relation between the thermodynamic potential F ($= pv - A$) and the entropy S.

According to **(1)**,

$$F = -U + pv - TS \qquad . \qquad . \qquad . \qquad (159)$$
$$dF = -dU + pdv + vdp - TdS - SdT.$$

Now,

$$dS = -\frac{dU + pdv}{T},$$

and therefore

$$dF = vdp - SdT, \left(\frac{\partial F}{\partial T}\right)_p = -S.$$

If we now introduce the heat function W_p for $-U + pv$, formula **(159)** becomes

$$F - W_p = T\left(\frac{\partial F}{\partial T}\right)_p,$$

an equation identical with **(158)**.

We can also arrive at the same result in the present case by a simple treatment, making use of a peculiarity of condensed systems which consists in the fact that with them we may operate even at the pressure $p = 0$. At low temperatures, where the vapour pressure is vanishingly small, this is immediately clear, but even at higher temperatures, in view of the smallness of the influence exerted by moderate changes of pressure on condensed systems, we may obviously, by a small extrapolation, indicate with any desired accuracy the behaviour of the condensed system which would be observed at zero pressure. At zero pressure, however, we must also perform the heating at constant pressure, because

* "Zeitschr. physik. Chemie," **62,** 119 (1908).

GENERALIZED TREATMENT

then no expenditure of work results from the heating associated with it. We thus obtain at once, most simply from a consideration of the appropriate cyclic process,

$$E_0 - W_0 = T\frac{dE_0}{dT} \quad \ldots \quad (p = 0) \quad . \quad (160)$$

or, generally,

$$K_n^0 - k_n^0 = T\left(\frac{\partial K_n^0}{\partial T}\right)_p \quad \ldots \quad (p = 0) \quad . \quad (161)$$

I have shown, in the paper just mentioned, how it is possible to proceed from the above equations to higher pressures by considering a cyclic process.

2. Formulation of the New Heat Theorem.—Equation (1) as applied to the expansion of a condensed substance by the volume dv being

$$A - U = T\left(\frac{dp}{dT}\right)_v, \quad . \quad . \quad . \quad (162)$$

the new Heat Theorem adds the condition,

$$\lim \left(\frac{dA}{dT}\right)_v = \lim \left(\frac{dp}{dT}\right)_v = 0, \text{ for } T = 0, \quad . \quad (163)$$

as we have already found on page 100.

Since, then, at low temperatures, according to the new Theorem,

$$\left(\frac{dp}{dT}\right)_v = 0 \text{ and } \left(\frac{dv}{dT}\right)_p = 0,$$

there is no need to distinguish here between

$$\left(\frac{\partial K_n}{\partial T}\right)_p \text{ and } \left(\frac{\partial K_n}{\partial T}\right)_v.$$

Discussions, which have appeared from time to time in the literature as to which of these quantities should converge towards zero at low temperatures, have therefore been founded on a misconception.

We now, therefore, formulate the new Heat Theorem in the statements

$$\lim \left(\frac{\partial K_n}{\partial T}\right)_p = \lim \left(\frac{\partial K_n}{\partial T}\right)_v = 0 \quad . \quad . \quad (164)$$

and thus also

$$\lim \left(\frac{\partial k_n}{\partial T}\right)_p = \lim \left(\frac{\partial k_n}{\partial T}\right)_v = 0 \qquad . \qquad . \quad (165)$$

3. Thermodynamic Potential.—The energy of transformation of condensed systems one into the other is, of course, accessible to calculation if we know their vapour tensions or dissociation tensions: in place of these the solubilities may also be employed.

Thus we have for the energy of transformation at constant pressure in the transition from monosymmetric to rhombic sulphur,

$$- F = RT \log_e \frac{\pi_1}{\pi_2}, \qquad . \qquad . \quad (166)$$

where π_1 and π_2 denote the vapour pressures of the two modifications: the equation refers to a mol of sulphur, as determined by Avogadro's law. The same formula holds for the reaction

$$2Ag + I_2 = 2AgI, \qquad . \qquad . \qquad . \quad (167)$$

π_1 being now the vapour pressure of solid iodine and π_2 the dissociation pressure of silver iodide. The Heat Theorem then gives

$$\lim \frac{dW}{dT} = \lim \frac{dF}{dT} = 0 \text{ (for } T = 0), \qquad . \quad (168)$$

where W denotes the heat of transformation at constant (atmospheric) pressure, obtainable from thermo-chemical measurements.

As frequently mentioned, the function A ($=$ maximum work) which we have used before, is practically identical with $- F$ for condensed systems if the pressures concerned are small; the electromotive force is an accurate measure of the chemical potential.

We have already learned numerous applications of equation (168) in earlier chapters, and it is natural that it should be possible to extend the new Heat Theorem very often to chemical and electro-chemical processes.

GENERALIZED TREATMENT

4. Surface Tension.—Both the work σ required for the formation of unit surface and the heat development q which is associated with the disappearance of unit surface (without doing work) must, at low temperatures, be equal to one another and independent of the temperature.

5. Coefficient of Magnetization.—As was shown by Warburg * in discussing some observations made by W. Thomson in 1878, the law for the magnetization of paramagnetic bodies under the influence of a field of intensity k is

$$A = km, \quad U = k(m - M), \quad (p = \text{const.})$$

Here m is the magnetic moment per unit mass produced by the magnetic force 1, and M is the heat absorbed in the process. The Second Law gives

$$M = T\left(\frac{dm}{dT}\right)_p \qquad . \qquad . \qquad . \quad \textbf{(169)}$$

and the new Heat Theorem supplies the further relation,

$$\lim \frac{M}{T} = \lim \left(\frac{dm}{dT}\right) = 0 \text{ (for } T = 0) \quad . \quad \textbf{(170)}$$

In agreement with this, Oosterhuis † showed that even substances which follow Curie's Law down to low temperatures deviate from it when the cooling is sufficiently intense, in a direction such that the susceptibility begins to be independent of temperature; he attempted to indicate an explanation of this behaviour based on molecular theory.

6. Thermo-electricity.—If we denote by Q the Peltier heat and by e the potential difference between two metals, we have the well-known equation (analogous to **(169)**)

$$Q = T\left(\frac{de}{dT}\right)_p ; \qquad . \qquad . \qquad . \quad \textbf{(171)}$$

the new Heat Theorem gives

$$\lim \frac{Q}{T} = \lim \left(\frac{de}{dT}\right) = 0 \text{ (for } T = 0) \quad . \quad \textbf{(172)}$$

---

* "Wied. Ann.," **20**, 814 (1883).
† "Physik. Zeitschr.," **14**, 862 (1913).

As Wietzel (86) has been able to show from an examination of a large number of combinations down to the temperature of hydrogen boiling under reduced pressure, there is an unmistakable tendency for $\frac{de}{dT}$ to fall off rapidly at low temperatures; in some exceptional cases this decrease can, however, only be observed at the lowest temperatures obtainable by the use of helium.* Equation **(172)** may be regarded as sufficiently verified by experiment.

7. Introduction of the Specific Heats.—The above formulæ are very much extended by introducing the specific heats; although we have learned numerous examples of this in previous chapters, we shall again deal briefly with their influence in the generalized formulæ.

The First Law supplies the relations

$$\left(\frac{\partial k_n}{\partial T}\right)_v dw_n = C_v' - C_v'', \qquad \left(\frac{\partial k_n}{\partial T}\right)_p dw_n = C_p' - C_p'' \quad (173)$$

where C_v' and C_p' denote the heat capacities of the system before the change of state has occurred which is defined by the change of the parameter w_n by dw_n, and C_v'' and C_p'' those after the change of state.

Integration of equation **(157)** gives

$$K_n = -T\int\frac{k_n}{T^2}dT + CT,$$

and the application of the new Heat Theorem,

$$\lim \frac{dK_n}{dT} = 0 \text{ (for } T = 0),$$

supplies the condition that the expression

$$-\int\frac{k_n}{T^2}dT - \frac{k_n}{T} + C$$

must disappear at low temperatures; this is only possible if

$$C = 0 \text{ and } \lim \frac{dk_n}{dT} = 0.$$

* Keesom, " Physik. Zeitschr.," **14,** 674 (1913).

GENERALIZED TREATMENT

In this case

$$K_n = -T\int^T \frac{k_n}{T^2}dT, \quad . \quad . \quad . \quad (174)$$

or, in other words, K_n may be calculated for all temperatures if we know k_n as a function of temperature; this is so if we know k_n at one given temperature and the values of $C_v' - C_v''$ or of $C_p' - C_p''$ for all temperatures.

8. Use of the T³-Law for Specific Heats.—In what follows we shall assume that Debye's relation,

$$E = aT^4$$

holds, at sufficiently low temperatures, not only for bodies under atmospheric pressure, but also for any pressures and for any modifications by magnetization, dielectric polarization, etc. It is true that there is not yet any experimental proof of this, nor will any direct proof be possible in many cases, but in all probability this assumption may be supposed to be of general validity, and the ideas of Debye may also, apparently, be generalized from their theoretical aspect.

Equation of State at Low Temperatures.—According to the Second Law we have in general *

$$p - \frac{\partial U}{\partial v} = -\frac{\partial Q}{\partial v} = T\frac{\partial p}{\partial T}; \quad . \quad . \quad (175)$$

$$T\frac{\partial^2 p}{\partial T^2} = \frac{\partial C_v}{\partial v} \quad . \quad . \quad . \quad (176)$$

We put

$$\frac{\partial E}{\partial T} = C_v = 4aT^3 \quad . \quad . \quad . \quad (177)$$

where a is a function of the volume only: to make this clear we shall write

$$C_v = 4f(v)T^3. \quad . \quad . \quad . \quad (178)$$

From **(176)** it therefore follows that

$$4f'(v)T^3 = T\frac{\partial^2 p}{\partial T^2}.$$

---

* " Theor. Chem.," pp. 60 and 61.

THE NEW HEAT THEOREM

By integration, taking note of the fact that

$$\lim \frac{\partial p}{\partial T} = 0 \text{ (for T = 0)},$$

it follows that

$$p = p_0 + \frac{f'(v)}{3}T^4, \qquad . \qquad . \qquad . \quad (179)$$

and hence

$$-\frac{\partial Q}{\partial v} = \frac{4f'(v)}{3}T^4 \qquad . \qquad . \qquad . \quad (180)$$

In order to obtain, finally, the thermal expansion of the body at zero pressure, we put

$$v_0 - v = \kappa p \qquad . \qquad . \qquad . \quad (181)$$

(κ being the compressibility at the volume v_0); then

$$\left(\frac{\partial p}{\partial T}\right)_v = \frac{\kappa \frac{\partial v_0}{\partial T} - (v_0 - v)\frac{\partial \kappa}{\partial T}}{\kappa^2} \qquad . \qquad . \quad (182)$$

and thus

$$\left(\frac{\partial p}{\partial T}\right)_{v=v_0} = \frac{1}{\kappa}\frac{\partial v_0}{\partial T} \qquad . \qquad . \qquad . \quad (183)$$

From **(179)** and **(181)** we then have

$$v_0 = V_0 + \kappa\frac{f'(V_0)}{3}T^4; \quad \left(\frac{\partial v_0}{\partial T}\right)_{p=0} = \kappa\frac{4f'(V_0)}{3}T^3 \quad (184)$$

Here V_0 signifies the volume at $T = 0$ and $p = 0$. Also, it follows from **(182)** that

$$\kappa = \kappa_0 + \beta T^4 \qquad . \qquad . \qquad . \quad (185)$$

The above equations show that the coefficient of expansion and the pressure-coefficient increase at low temperatures proportionally to the third power of the absolute temperature; the heat of compression varies as the fourth power.

The fact that thermal expansion and specific heat are proportional to one another, on certain assumptions, even at higher temperatures, was, of course, found by Grüneisen,*

* "Verhandl. D. physik. Ges.," **13**, 426 (1911).

GENERALIZED TREATMENT

but it is only at low temperatures that there is any question of a rigid law.

For the influence of the specific heats on the chemical affinity on the assumption of the T^3-law, cf. page 82.

Surface Tension—The quantities σ and q mentioned on page 217 are related to one another by the formulæ

$$\sigma = q_0 + \frac{a}{3}T^4, \qquad q = q_0 - aT^4;$$

the coefficient a is given by the change in thermal capacity which is associated with the formation of unit surface.

Coefficient of Magnetization.—Here we have (at constant pressure)

$$m = m_0 + \frac{a}{3}T^4$$
$$m - M = m_0 - aT^4$$
$$M = \frac{4}{3}aT^4.$$

The coefficient a is determined by the change in thermal capacity, which is experienced by unit volume of a paramagnetic substance when magnetized by the force $k = 1$; we have

$$\frac{d(m - M)}{dT} = 4aT^3 = \tau_p' - \tau_p''.$$

τ_p' is the thermal capacity of unit volume before, and τ_p'' that after the magnetization. As $\tau_p' - \tau_p''$ is of course inaccessible to direct measurement owing to its extreme smallness, the foregoing equation may serve for an indirect derivation of this quantity, which may perhaps become of significance for the theory of paramagnetism.

Peltier Heat.—In this case (cf. p. 217)

$$e = e_0 + \frac{a}{3}T^4, \quad \frac{de}{dT} = \frac{4}{3}aT^3, \quad Q = e - U = \frac{4}{3}aT^4 \quad \textbf{(186)}$$

I have already had occasion [*] to point out that the Peltier effect divided by the absolute temperature must be infinitely

[*] "Sitzungsber. d. Preuss. Akad. d. Wiss.," 1912, p. 140.

small to the same order of infinity as the specific heats. Keesom * has arrived at the above formulæ by a totally different method.

We have here the peculiar condition that we are not in a position to make a reliable determination of the coefficient *a* from specific heats by even an imaginary experiment.

9. Influence of Temperature on Gravitation.—The " actions at a distance " of classical physics have hitherto, as is well known, refused to be subject to the application of the Second Law : all thermodynamical questions appeared to be conditioned by the hypothesis (cf. p. 10),

$$A = U.$$

The method of thermodynamical treatment indicated by our Heat Theorem once more takes us a step further. The above equation cannot possibly hold for all temperatures, for it assumes that a body has exactly the same thermal capacity whether it is isolated from the influence of the particular distant force or is subjected to it. This may be very nearly the case, but it cannot possibly be exactly so, and thus it may happen that both A and U may be, at low temperatures, practically but not fundamentally independent of temperature and thus equal to one another.

In order to make as clear as possible this line of reasoning, which seems to me to be of deep significance, let us put it into other words. The fall of a stone to the earth is a process which takes place at ordinary temperatures (and certainly also at higher temperatures) exactly in the same way as, say, the silver-iodine cell described on page 113 would behave at 1° abs. ; but in both cases it is only a question of raising the temperature in order to be able to observe an appreciable divergence between A and U.

We shall follow this out by calculation for the gravitational case. For this purpose we shall assume that there is

---

* " Physik. Zeitschr.," **14**, 670 (1913).

GENERALIZED TREATMENT

a point of mass m inside a large hollow sphere of constant temperature, and that it is first removed from the influence of gravitation and then brought into a gravitational field, under the influence of which it can take up the velocity v.

The hypothesis hitherto made then states simply

$$A = U = \frac{m}{2}v^2, \qquad \qquad (187)$$

and we must now write

$$U = \frac{m}{2}v_0^2 + E - E', \qquad \qquad (188)$$

$$A = \frac{m}{2}v_0^2 - T\int_0^T \frac{E - E'}{T^2} dT, \qquad \qquad (189)$$

where E denotes the energy content of the massive point (at rest) before, and E^1 that after the above described process. At low temperatures **(188)** and **(189)** reduce, of course, to **(187)**.

In order to obtain an insight into the magnitude of $E - E'$, and thus of the temperature-effect, we shall consider an atom of hydrogen as a massive point. We cannot, of course, do this without some hypothetical assumptions, but if the idea is correct, which has been so frequently expressed in recent times, that all matter is composed of positive electrons (= hydrogen ions) and negative electrons, the above specification will always be convenient for any future theory of gravitation, since it represents the simplest case, to which all the phenomena of gravitation may be reduced.

The action of gravity on the hydrogen atom extends, of course, almost exclusively to the nucleus; we shall neglect the action on the negative electron. For the constitution of the nucleus we make use of the idea which I have developed in my paper (108) (pp. 109 *et seq.*), according to which we may regard it as a resonator with the frequency of oscillation, or, more correctly, of revolution,

$$\nu = 0 \cdot 75 \times 10^{23}.$$

Then we have

$$E = \frac{h\nu}{e^{\frac{h\nu}{kT}} - 1}, \qquad E' = \frac{h\nu'}{e^{\frac{h\nu'}{kT}} - 1},$$

where there is between ν and ν' the simple relation

$$\frac{\nu - \nu'}{\nu} = \left(\frac{v_0}{c}\right)^2 \qquad . \qquad . \qquad . \qquad (190)$$

(c is the velocity of light and v_0 the value of v reduced to $T = 0$). The equations **(188)** and **(189)** given by our Heat Theorem then assume the form,

$$U = \frac{m}{2}v_0^2 + \frac{h\nu}{e^{\frac{h\nu}{kT}} - 1} - \frac{h\nu'}{e^{\frac{h\nu'}{kT}} - 1}, \qquad . \qquad . \qquad (191)$$

$$A = \frac{m}{2}v_0^2 - h(\nu - \nu') + kT \log_e \frac{e^{\frac{h\nu}{kT}} - 1}{e^{\frac{h\nu'}{kT}} - 1} \qquad . \qquad (192)$$

so that our problem is solved, albeit by making use of what are at present very hypothetical assumptions. If we limit ourselves to temperatures which are low compared with $\frac{h\nu}{k} = 3.7 \times 10^{12}$ degrees, equations **(191)** and **(192)** reduce, by making use of **(190)**, to

$$U = \frac{m}{2}v_0^2 - h\nu \frac{\left(\frac{h\nu}{kT} - 1\right)\left(\frac{v_0}{c}\right)^2}{e^{\frac{h\nu}{kT}}}, \qquad . \qquad . \qquad (193)$$

$$A = \frac{m}{2}v_0^2 + h\nu \frac{\left(\frac{v_0}{c}\right)}{e^{\frac{h\nu}{kT}}} \qquad . \qquad . \qquad . \qquad (194)$$

Formula **(187)** thus continues to hold at temperatures which are small compared with 3·7 billions of degrees, but we must remember the limitation that, strictly, it is only very near the absolute zero that the heat developed by the arrest of a falling stone is equal to the maximum work of the process.

10. Classification of Natural Processes.—We have already learned (on p. 10) to know two special categories of natural processes distinguished by the limiting cases,

$$U = A \text{ and } U = 0.$$

The former is the more general in that all processes tend towards this limit at sufficiently low temperatures.

We can now go a step further and suppose a more universal classification by taking note of the temperatures at which the processes in question begin to depart from the limiting region, where

$$U = A.$$

Our Heat Theorem has taught us to recognize a criterion for this in the oscillation frequency ν, which, multiplied by $\beta = \dfrac{h}{k}$, gives this characteristic temperature. We are concerned here, of course, only with a rough approximation, and the more accurate result has to be determined in each case as explained in this book. It is therefore immaterial in this connection whether we determine the characteristic frequency of oscillation with the formula of Einstein or with that of Debye. With this limitation, the following list of characteristic absolute temperatures is perhaps worthy of note :—

Process.	Characteristic Temperature.
Expansion of a gas of normal concentration (1 mol per 22.41 litres)	0·03°–0·0001°
Chemical and electro-chemical processes in condensed systems, as a rule between	100°–500°
Radio-active processes	ca. 5×10^{10} degrees.
Gravitational phenomena	ca. 4×10^{12} ,,

The number given for radio-active processes has been calculated by assuming that the kinetic energy of the helium atoms emitted, which is of the same order,* e.g. for RaC, RaF, Actin-B, Thorium-B and -C is equal to $h\nu$. There is no means of guessing the effect of temperature on the

---

* Cf. Kohlrausch, " Leitfaden d. prakt. Physik," 11 Aufl., S. 642.

Coulomb force of attraction ; the hypothetical character of the last number may be once more emphasized.

11. Recapitulation.—In concluding our account of the thermodynamical questions which have occupied our laboratory for the last decade, the question arises, when the results obtained are reviewed :—

Has thermodynamics now reached finality, or is the discovery of other Heat Theorems to be expected ?

I think this question may be answered quite definitely and certainly by saying that if, as I hope to have shown, our Heat Theorem is really to be regarded as a third to the two existing Laws in the universality of its validity and accuracy, there cannot be a fourth or any other main law.

In fact, thermodynamics seems to be exhausted by the addition of our Heat Theorem to the two existing Laws ; for all processes of nature can now be definitely calculated from a knowledge of the energy function U, in so far as this is possible in view of the fact that the co-ordinates of space and time cannot be included in thermodynamics ; all the further advances which are conceivable in the field of thermodynamics can therefore relate only to further regularities governing that function. But U may be referred to the energy U_0 at the absolute zero and to specific heats, both of which quantities are in their nature foreign to general thermodynamics.

We have thus a clear idea of the problems of the future ; their solution will hardly advance thermodynamics at all in principle, but so much the more will they facilitate its application. It will be necessary, first, to explore further the dependence of specific heats on temperature and all conceivable external influences and, second, to obtain a clearer idea of the two identical quantities U_0 and A_0 (energy at the absolute zero and potential energy) ; in the latter direction my paper (108) already cited may constitute, perhaps, a first attempt.

CHAPTER XVI

SOME HISTORICAL AND MATERIAL ADDENDA

I HAVE limited myself in the foregoing chapters mainly, though by no means exclusively, to a discussion of the results arrived at by my co-workers and myself; it will be desirable, for the sake of completeness, to deal briefly in this supplementary chapter with the work of some other authors.

1. Early History of the Heat Theorem.—I have already been at some pains, in the first chapter, to describe the evolution of the new Heat Theorem historically, i.e. indicating those papers to which I owed important suggestions in my long preparatory work.

I may here refer particularly once more to the well-known and excellent text-book of Haber, " Thermodynamik technischer Gasreaktionen " (1905), which, however, did not appear until the essential part of my early work was concluded. In this book particular emphasis is laid on the importance of the integration constants in gaseous equilibria; the nature of these our Heat Theorem then explained.

I have intentionally omitted to mention a paper by Th. W. Richards,* because I cannot admit that the thermodynamics in it had any influence on my deliberations owing to its vagueness and incorrectness; van't Hoff remarks, however, that he was prompted by the above-mentioned publication of Richards to his work, with which I have dealt on page 9, which, though not unexceptionable in principle, was yet prolific in suggestions.

* " Zeitsch. physik. Chem.," **42**, 129, 1902.

I discussed this paper on page 56 of my Silliman lecture as follows: " In 1902 a very interesting paper was published by Th. W. Richards on ' The Relation of changing Heat Capacity to Change of Free Energy, Heat of Reaction, Change of Volume, and Chemical Affinity,' in which he pointed out very clearly that the question whether $A > U$ or $U < A$ above absolute zero (where $A = U$) depends upon whether the heat capacity is increased or decreased by the chemical process, and I am very glad to be able to state that our formulæ agree qualitatively in many cases with the conclusions of Richards. I do not wish to enter here into a discussion of the differences in the quantitative relations."

I thought that no one reading these lines carefully could mistake my meaning, namely, that, in spite of the many suggestive ideas of Richards' paper, either his or my thermodynamical theorems must be wrong, for I emphasized the partial character of the qualitative, and the absence of quantitative, agreement; thermodynamics is, of course, specially distinguished by its definite and precise conclusions.

Now Richards imagines of late * that I had here admitted the identity of his older conception with my more recent one. From this it seems to me to be evident that Richards has understood neither the meaning of the above lines nor my Heat Theorem correctly. He even adds: " All these ideas were afterwards (1906) adopted unchanged by Nernst in his development of the ' Wärmetheorem ' usually named after him." I am thus obliged to enter into a criticism of Richards' work. I have hitherto regarded it as unnecessary, in view of the clearness of the position, and thought that I might be able to limit myself to the above remarks in order to avoid superfluous polemics. I am, however, obliged to refute the claim made by Richards to my Heat Theorem.

In order to answer the question of the relation between maximum work A and heat evolution U, Richards dispenses

---

* " Journ. Amer. Chem. Soc.," **36**, 2433 (1914).

HISTORICAL AND MATERIAL ADDENDA

with the application of the Second Law, and thinks to make the First Law suffice, a procedure which will of itself startle any one who is acquainted with classical thermodynamics. By considering a cyclic process which he seems to me to describe very vaguely, Richards obtains with the aid solely of the First Law, the relation,

$$U = A + E - E' \quad . \quad . \quad . \quad (R\,!)$$

(E being the heat content before, and E' that after the change). But this equation is of course wrong; correctly it must read (Kirchoff's Law)

$$U = U_0 + E - E',$$

where U_0 is the heat evolution at the absolute zero. According to the Second Law, A_0 may be substituted for U_0.

I may spare the reader of this book the recital of a number of other thermodynamical oddities and vague results at which Richards arrives by means of his formula $(R\,!)$; they lead him eventually to consider that the change of free energy is not, in general, the measure of the chemical affinity: " If the heat capacity diminishes during the reaction the free energy is less than the affinity and *vice versa* " (*sic*).

On the ground of the above paper, Richards apparently considers himself the real discoverer of my Heat Theorem, although his formulæ and mine have not the least thing in common. Even if, as seems hardly to be doubted, he has not understood the meaning of my equations at all, the following circumstance should have given him food for thought.

Van't Hoff, in seeking to give even a moderately clear form to Richards' work, arrived in 1904, in his paper mentioned on page 9, at the result

$$\lim \frac{dA}{dT} = \pm \infty \text{ for } (T = 0),$$

while my Heat Theorem, on the contrary, states that

$$\lim \frac{d\mathrm{A}}{d\mathrm{T}} = 0 \text{ (for T = 0)}.$$

It may be pointed out, further, that Richards has never raised any objection against van't Hoff's result.

Again, in 1914 Richards (*loc. cit.*) regards it as certain that, provided concentration effects are avoided, A is greater than U if the heat capacity of the reacting substances is greater after the reaction than before and *vice versa*, in agreement with his rule, which I have already mentioned above. Here, too, Richards is in error. This rule is obeyed very often, it is true, but by no means always. The rule would mean that if U increases with the temperature the relation A < U must hold and *vice versa*. But if we consider the diagram of Fig. 16 (p. 112), we see that though the rule does hold up to about T = 125°, it does not continue to do so above this temperature. In this case there can be hardly any doubt of the reliability of the measurements.

To mention a second very simple example, U is certainly greater than A in the transformation of quartz glass into crystallized quartz. In accord with Richards' rule, the specific heat of quartz glass at low temperatures is therefore notably greater than that of quartz. At higher temperatures, very definitely even at T = 400°, the converse is, however, the case, and at T = 600° the difference is fairly considerable (cf. paper 47, p. 430). The fact that quartz glass has a lower specific heat than quartz above room temperatures has also been confirmed recently by the careful experiments of K. Schulz.* At about room temperature the specific heats of the two modifications are equal, so that here, according to Richards, there must be equality between A and U, which is of course out of the question.

All these matters may be regarded extremely simply from the standpoint of my Heat Theorem. At low tem-

---

* "Zentralbl. f. Mineralogie und Geologie" (1912), 481.

peratures the A and U curves must always deviate from one another, hence here Richards' rule holds. At higher temperatures this is usually still the case, but occasionally, owing to variation of specific heat due to circumstances of a more or less secondary character, the U curve attains a maximum or minimum and then bends round; then Richards' rule no longer holds, so that it has only a quite subsidiary significance. Richards therefore completely mistakes the position when he raises even distant claims to my Heat Theorem on the grounds of this rule.

2. Some Further Applications of the Heat Theorem to Condensed Systems.—The reaction discussed on page 113 has been examined in a paper by G. Jones and G. L. Hartmann,* though not with the completeness with which U. Fischer and, later, Braune and Koref (p. 115), had done so. These authors limited themselves to a measurement of the temperature-coefficient of the cell,

$$Ag \mid AgI \mid I_2;$$

with the aid of the formula of Helmholtz they calculate for $T = 298$

$$U = 14{,}570 \text{ cals.}$$

They state that the purpose of their work (carried out in Richards' laboratory) was to test Richards' theory of " compressible atoms." They rightly point out that the above value is some 500 cals. less than is required by my Heat Theorem (15,014, according to p. 115).

As regards the deviation of their result from that of Fischer, they emphasize the fact that their cells were more constant; they omit to mention that Fischer has verified his value in a totally different manner by careful direct thermo-chemical measurements, made by two different methods; nor do they refer to the results of Braune and Koref, who were able to confirm Fischer's number very

* " Amer. Chem. Soc.," **37**, 752, 1915.

closely by direct determination of the heat change. In view of the far-reaching conclusions which Jones and Hartmann draw they ought not, in my opinion, to have omitted to make a direct thermo-chemical check on their value of U.*

Shortly afterwards H. S. Taylor † submitted the above work to what seems to me a very fundamental and reasoned criticism, and brought forward fresh experimental material, in particular from an examination of the cell,

$$Pb \mid PbI_2 \mid AgI \mid Ag.$$

From the temperature coefficient of this cell he derives the value

$$U = 11,550 \text{ (calc. } 11,723);$$

the number included in the brackets results from an application of my Heat Theorem, and is thus in agreement within the error of observation. If the same value is deduced from the determinations, carried out by Braune and Koref, of the heat of formation of lead iodide, we obtain

$$U = 11,650;$$

making use of the heat of formation of silver iodide, obtained by Fischer by electrometric means, and of the Braune-Koref value for lead iodide, the value obtained is

$$U = 11,620.$$

Combination of the heat of formation of silver iodide obtained by Jones and Hartmann with the value of Braune and Koref for lead iodide gives

$$U = 12,830.$$

In view of the fact that the figure of Braune and Koref for lead iodide was obtained in excellent agreement (cf. p. 118) by two totally different methods, electrometric and thermochemical, the discrepancy shown by the last-mentioned value

---

* See also Supplement, p. 270.
† "Amer. Chem. Soc.," **38**, 2295 (1916).

is a further strong argument for the exactness of Fischer's work and against the accuracy of the measurement of Jones and Hartmann. In opposition to these authors, Taylor, in his conclusion, regards the results "as in favour of the Nernst Theorem within the present accuracy of measurement of specific heats."

We are indebted to the same author [*] in conjunction with two co-workers for an examination of the two cells,

$$Cd \mid CdCl_2, 2^{1/2}H_2O \mid HgCl_2 \mid Hg,$$
$$Cd \mid CdSO_4, {}^{8/2}H_2O \mid Hg_2SO_4 \mid Hg,$$

from the standpoint of my Heat Theorem. The calculations were carried out in a manner quite analogous to the treatment of the Clark cell by Pollitzer (p. 116). The values of U at room temperatures were obtained from the temperature-coefficients of the two cells which are particularly well known; fresh measurements were made of the specific heats of the two cadmium salts (cf. also the following section). The result obtained for $T = 291$ was for the former cell,

0·7236 volts (found), 0·7426 volts (calc.),

and for the latter (Weston cell with pure cadmium instead of amalgam)

1·072 volts (found), 1·071 volts (calc.).

The agreement must be regarded as extremely satisfactory, particularly as the specific heats of the two cadmium salts were only measured down to the temperature of liquid air. A more detailed discussion appears to be desirable of the question how the small temperature coefficient of the Weston cell compared with that of the Clark cell is to be explained, and how, in a more accurate calculation, account is to be taken of the heat of dilution at the cryohydric point. The cryhohydric point for the Weston cell was found to be $T = 256°$ (cf. also my remarks (p. 118), and Landolt and Börnstein, "Tabellen," p. 464, IV Aufl.).

[*] F. M. Seibert, G. A. Hulett, H. S. Taylor, "Amer. Chem. Soc.," **39**, 38 (1917).

3. Measurement of Specific Heats.

—In amplification of my remarks on pp. 50-52, reference may be made also to the investigation of Keesom and Kamerlingh-Onnes * on the specific heat of copper between 14° and 90° (abs.). The result here obtained is that the values of C_v may be represented by means of Debye's formula with the value $\beta\nu = 315$, with a concordance nearly as good as the accuracy of the measurements; this value of $\beta\nu$ is identical with that which I had derived from my older and of course less accurate measurements (cf. p. 52).

Reference may also be made here to the new measurements of specific heat by means of the vacuum calorimeter, to which we have already alluded on page 233. The authors mentioned made use of an apparatus which was essentially the same as that described by Schwers and myself; the temperature was measured by means of a platinum wire, through which passed an extremely constant current, so that the resistance could be measured by the potential drop, using a potentiometer. During the heating a silver voltameter was included in the circuit as a control on the measurement of time; convenient though this is, I think that as a rule the determination of the duration of the heating current can be made with sufficient accuracy to allow the somewhat troublesome silver estimation to be avoided.

4. Practical Application of the Heat Theorem.

—As a practical rule for the application of our Heat Theorem we have made use on page 81 of a method of expressing the equations

$$A - U = T\frac{dA}{dT} \quad . \quad . \quad . \quad (1)$$

$$\text{and } \lim \frac{dA}{dT} = 0 \text{ (for } T = 0) \quad . \quad . \quad (35)$$

in the shorter form

$$A = -T\int^T \frac{U}{T^2}dT. \quad . \quad . \quad (39)$$

---

* "Sitzung. der Kgl. Akademie zu Amsterdam," 25 Sept., 1915.

HISTORICAL AND MATERIAL ADDENDA

It was pointed out that the expression following the sign of integration was to be integrated indefinitely, and the upper limit to be introduced into the expression so obtained.

It has been objected by a very eminent mathematician that this method of writing it is not compatible with the nature of integration; it is quite easy to put the matter right, but, on the other hand, the rule for calculation given by formula **(39)** is perfectly correct in its operation, so that it appears to me to be not unsuitable. It may readily be seen from a simple example, such as formula **(41)**, that no lower limit can be indicated for the integral of equation **(39)**.

As Professor Byk was kind enough to inform me, and as may readily be checked, the expression *

$$A = U - T \int_0^T \frac{dU}{dT} \cdot \frac{dT}{T}$$

also satisfies the above pair of formulæ, **(1)** and **(35)**; in place of U we have here after the integral only the specific heats (cf. formula **29**). But if the complete A-curve is to be obtained from the U-curve, the application of equation **(39)** will as a rule be preferable.

In 1915 Gans and Pereyra Miguez,† on the one hand, and Drägert ‡ on the other, independently and simultaneously described an apparatus which depends on the construction described on page 84, and draws the A-curve automatically, when the U-curve is traced from the absolute zero with the pointer of the apparatus. Particular attention may be drawn to this very ingeniously contrived thermodynamical integrator.

5. The Heat Theorem and the Quantum Theory.—We have seen in Chapter VII that the new Heat Theorem may

---

* Cf. also Planck, "Thermodynamik," 5 Aufl., p. 283 (formula 282).

† "Physik. Zeitsch.," **16**, 247 (1915).

‡ *Ibid.*, **16**, 295 (1915); cf. also paper (102).

be derived from the two existing Laws of Thermodynamics on the assumption that the specific heats at low temperatures are equal to zero of at least the first order. This result follows from the Quantum Theory with the exception that the zero which is reached is one of a considerably higher order; this is the case not only for solids, but also for gases of finite concentration. We have here, therefore, an indication that there is a close relationship between the Heat Theorem and the Quantum Theory or, more correctly, that group of phenomena which Planck has by his Quantum Theory included in a common point of view with such successful results. In my lecture, "Neuere Probleme der Wärmetheorie" (48), I have referred to this in more detail; Sackur * and Jüttner † have mentioned special considerations which, on the basis of probabilities lead from the Quantum Theory to our Heat Theorem.

I may here content myself with drawing attention to the following quite general point of view. The fundamental idea of the quantum theory leads to the conception that every homogeneous system (gas, gaseous mixture, crystal, solution, etc.) must, on cooling, reach a state in which, even before the absolute zero is attained, further cooling exerts absolutely no recognizable influence on any property of the system concerned. Recent measurements have in some cases proved, and in others rendered it probable, that this is true for the volume, the compressibility, the electrical resistance, the thermo-electric activity, the magnetic susceptibility, etc. [cf. Nernst (62a)], and in particular for the thermodynamic functions U and A. Our Heat Theorem,

$$\lim \frac{d\text{U}}{d\text{T}} = \lim \frac{d\text{A}}{d\text{T}} = 0 \text{ (for T = 0)},$$

is thus, as it were, a special case of a much more general law; it is true that, as we have seen in Chapter XV, from this

---

* "Ann. d. Phys." (4), **34,** 465 (1911).
† "Zeitsch. f. Elektrochemie," **17,** 139 (1911).

HISTORICAL AND MATERIAL ADDENDA 237

special case may be deduced not only the other cases, but also many other important peculiarities.

A very important application of the quantum theory to the energy content of a substance at very high pressures is due to M. Polányi,* who arrives by it at the result that at very high pressures the specific heats must always converge towards zero.

We may illustrate this result by the following simple treatment. As Madelung and Einstein have shown,† the frequency ν of atomic vibration may be calculated from the compressibility ; if this formula is applied to a body which is under very high pressure, and of which the compressibility under these conditions has become very small, the values of ν must obviously become extremely high and, conversely, the specific heat must become very small at temperatures which are not particularly high. Any theory of the "degeneration" of gases (p. 193) leads to the conception that a gas very strongly compressed above the critical point has a very low heat capacity.

For processes under very high pressure—and this is the particularly noteworthy point in Polányi's treatment—we have, even at finite temperatures,

$$\frac{dA}{dT} = \frac{dU}{dT} = 0,$$

or, in other words, the length of the section which starts from the absolute zero and is common to the A- and U-curves, is increased at high pressures. We cannot here deal with other very interesting results which Polányi deduces from his considerations.

6. Some Questions of Principle.—In Chapter VII I have reproduced and stated in rather more detail the proof which I gave as early as 1912, whereby my Heat Theorem may be derived from classical thermodynamics if the disappearance

* " Verhandl. d. D. Physik. Ges.," **15**, 156 (1913).
† Cf. " Theor. Chem.," p. 283.

of the specific heat at the absolute zero is assumed as an experimentally demonstrated fact.

The importance of this proof must not, however, be over-estimated ; I thought, in my first papers, that it was desirable to be cautious in upholding the idea that all atomic heats converge at the absolute zero towards a limiting value, which might be small but was characteristic of each element or, perhaps even, common to all elements. The idea is, however, quite reconcilable with my Heat Theorem, provided only that this limiting value is independent of the nature of the complex in which the element is present. This idea can never be refuted experimentally, since experiment gives only upper limits for these " residual atomic heats " ; if these are finite, however small they may be, my proof falls through, and so of course does Planck's hypothesis of the entropy becoming zero ; but there will not be the slightest alteration in the practical applications of my Heat Theorem.

In these circumstances, which so far as I can see are not challenged from any authoritative source, it appears that the original concept of my Heat Theorem,

$$\lim \frac{d\mathrm{U}}{d\mathrm{T}} = \lim \frac{d\mathrm{A}}{d\mathrm{T}} = 0 \text{ (for T = 0)},$$

(U and A being the changes of total and of free energy associated with the process under examination) is still to be preferred at the present time, because it does not burden with a superfluous hypothesis the calculations which are to be carried out with the Heat Theorem.

The proof which I have given above has been discussed by H. A. Lorentz, Czukor, Pólányi, and Einstein. H. A. Lorentz * limits himself to the communication of a deduction by which my Heat Theorem can be proved if the principle of the Unattainability of the Absolute Zero which I proposed is assumed to be correct. I myself was, and still am, of the opinion that this principle follows of necessity

---

* " Chem. Weekblad.," **10**, p. 621 (1913).

HISTORICAL AND MATERIAL ADDENDA

from the Second Law, if it be assumed that the specific heats become zero.

Czukor * considers a cyclic process, in the same way as I have done, and the difference from my derivation appears to me purely formal. The conception which he proposes for the Heat Theorem is, however, suggestive in many respects : " There is for every condensed substance a range of temperature in which reactions take place along an isothermal just as they would occur adiabatically at the ordinary temperature."

Polányi † treats the question in a somewhat different manner by considering a series of adiabatic and isothermal changes ; although his procedure is unexceptionable, I rather prefer the treatment of a simple cyclic process. In a second paper, Polányi ‡ chiefly maintains the cogency of his arguments against Einstein.

It cannot fail to be of great interest if I deal briefly, in conclusion, with Einstein's standpoint. At the first Solvay Congress § Einstein stated (1911) that my Heat Theorem could not be deduced from the disappearance of the specific heats near the absolute zero, and H. A. Lorentz supported his arguments. I showed, in reply, that near the absolute zero the ordinary potential theory was to be applied to any dislocation of an atom from the standpoint of the quantum theory ; this is, of course, identical with my Heat Theorem.

Soon afterwards I published the purely thermodynamical proof of my Heat Theorem, which has been reproduced and elaborated in a few respects in Chapter VII. Einstein did not at first regard this as conclusive, as may be seen from the following remarks,‖ which occur in the beginning of his paper :—

" Two considerations are reproduced below which are

---

* " Verhandl. d. D. Physik. Ges.," **16**, 846 (1914).
† *Ibid.*, **16**, 333 (1914).   ‡ *Ibid.*, **17**, 350 (1915).
§ " Deutsche Ausgabe," pp. 243 and 244.
‖ " Verhandl. d. D. Physik. Ges.," **16**, 820 (1915).

related in that they show how far the most important recent results of the theory of heat, viz., Planck's radiation formula and Nernst's Theorem, can be deduced, without the aid of Boltzmann's principle, by purely thermodynamical means by making use of the fundamental idea of the quantum theory. In so far as the conclusions given below correspond with the facts, Nernst's Theorem holds for chemically pure crystalline substances, but not for solid solutions. Nothing definite can be stated as regards amorphous substances, owing to the prevailing ignorance as to the nature of the amorphous state.

"In justification of the present attempt to deduce Nernst's Theorem theoretically, I must point out, in introduction, that all endeavours to do so by thermodynamical means, using the experimental rule that heat capacity vanishes at $T = 0$, are to be regarded as having failed. . . ."

Nevertheless, Einstein has in this paper approached very much nearer to my standpoint, in that he now deduces my Heat Theorem from considerations of quantum theory, at any rate for pure crystalline substances.

As concerns solid solutions (e.g. in the case of dilute solutions) it is of course clear that, if the gas laws are regarded as valid here down to the lowest temperatures, my Heat Theorem cannot hold, for the same reasons as it does not hold for gases in the corresponding case (cf. p. 191). If the theory of the "degeneration" of gases is taken as a basis there can be no doubt that, since the Heat Theorem is then true for gaseous mixtures, it must hold also *a fortiori* for solid solutions.

Another important advance is represented by a paper by O. Stern.[*] In the light of a very original method of treatment, Stern here comes to the result that two solid solutions of the same chemical composition, but having different arrangements of the components, can be transformed one

---

[*] "Ann. d. Physik" (4), **49**, 823 (1916).

into the other at low temperatures according to the equation

$$A = U.$$

This result is merely a special case of the idea, which I defended at the first Solvay Congress and have already mentioned, according to which any such alteration must, at low temperatures, take place according to the laws of the classical potential theory.

According to a private communication which he has been kind enough to make to me, Einstein accepts the proof brought forward by Stern. According, again, to a verbal communication, he now considers it very improbable that the adiabatic curve can be such as to intersect the axis of temperature (cf. pp. 88 and 89); we have now, therefore, reached complete agreement as to the force of the arguments which I have given in Chapter VII (or paper 65).

APPENDIX

1. Definitions of Symbols and Numerical Values.—In the absence of any remark to the contrary, the symbols given below have the following meaning:—

P, p	. .	Pressure.
V, v	. .	Volume.
T	. .	Absolute temperature.
π_0	. .	Critical pressure.
θ_0	. .	Critical temperature.
K	. .	Constant of the Law of Mass Action.
C_v	. .	Molecular heat at constant volume.
C_p	. .	Molecular heat at constant pressure.
E	. .	Heat content $\left(=\int_0^T C_v dT \text{ or } \int_0^T C_p dT\right)$.
F	. .	Content of free energy $\left(=T\int_0^T \frac{E}{T^2} dT\right)$.
ν	. .	Frequency of oscillation or of revolution.
$\beta\nu = \dfrac{h\nu}{k}$	. .	Specific temperature.

In certain places E also denotes the electromotive force of a galvanic cell, and EF the work supplied by it per electrochemical gramme-equivalent; on pages 214-216 F signifies the thermo-dynamical potential; no confusion is likely to be caused by these extra significations.

In the fundamental equation

$$pv = RT$$

the gas constant R is per mol; at N.T.P. (0° on the ordinary scale, or T = 273·09 on the absolute scale, and at atmospheric pressure) a mol of an ideal gas occupies 22·412 litres. Hence we have

$$R = 0\cdot0819 \left[\frac{\text{litre-atm.}}{°C}\right]$$

or

$$R = 8\cdot 315 \times 10^7 \left[\frac{\text{ergs}}{°C}\right]$$

or, finally, if, as we have practically always done in this book, we select the 15° gramme-calorie as the unit of energy,

$$R = 1\cdot 985 \left[\frac{\text{gm.-cal.}}{°C}\right]$$

Further, we use throughout

$\log_{10} x$ for the common logarithm of x,
$\log_e x$ for the natural logarithm of x.

For the two fundamental thermodynamical functions (total and free energy) U and A are always used; U signifies the heat developed in a process, or, in other words, the total amount of heat which can be measured in a calorimeter, and A the corresponding maximum work supplied in isothermal processes. This method of description, which frequently differs in sign from that hitherto customary, seemed to me advantageous for the following reasons :—

1. In thermo-chemistry the quantities of heat developed are always considered as positive.

2. On the above definition

$$dA = p\, dv,$$

i.e. dA is positive; the two fundamental equations **(1)** and **(43)**, cf. page 90,

$$A - U = T\frac{dA}{dT} \quad . \quad . \quad . \quad . \quad (1)$$

$$p - \frac{\partial U}{\partial v} = T\frac{dp}{dT} \quad . \quad . \quad . \quad . \quad (43)$$

are homogeneous as regards sign.

3. It has long been the rule in chemistry to write the equations of reactions as they normally occur, e.g.,

$$2H_2 + O_2 = 2H_2O,$$

and only in exceptional cases

$$2H_2O = 2H_2 + O_2.$$

Using the former method of writing the equation, U and A are, as a rule, positive if the choice of sign which we have used is accepted.

APPENDIX

In conclusion, the following numerical values of natural constants * may be tabulated, expressed in every case on the C.G.S. system:—

Velocity of light . . . $c = 3.00 \times 10^{10}$.
Planck's constant . . . $h = 6.55 \times 10^{-27}$.
Gas constant . . . . $R = 8.315 \times 10^{7}$.
Number of molecules per mol . $N = 6.17 \times 10^{23}$.
Mass of the hydrogen atom . $m = 1.63 \times 10^{-24}$.
Mass of the negative electron . $\mu = 8.9 \times 10^{-28}$.

$$k = \frac{R}{N} = 1.347 \times 10^{-16}; \quad \beta = \frac{h}{k} = 4.863 \times 10^{-11}.$$

2. Tables.—On the following pages will be found a number of tables which are very useful in calculations relating to the Heat Theorem:—

Table I. C_v after Einstein (cf. p. 59, formula **12**).
Table II. C_v after Debye (cf. p. 60).†
Table III. $\frac{E}{T}$ after Einstein (cf. p. 94).

$$\frac{E}{T} = 3R \frac{\frac{\beta\nu}{T}}{e^{\frac{\beta\nu}{T}} - 1}.$$

Table IV. $\frac{F}{T}$ after Einstein (cf. p. 95).

$$\frac{F}{T} = -3R \log_e \left(e^{\frac{\beta\nu}{T}} - 1\right) + 3R\frac{\beta\nu}{T}.$$

Table V. $\frac{E}{T}$ after Debye (cf. p. 95, formula **55**).

$$\frac{E}{T} = \frac{9}{12}R \left(\frac{C}{C_\infty} + \frac{3\frac{\beta\nu}{T}}{e^{\frac{\beta\nu}{T}} - 1}\right).$$

Table VI. $\frac{F}{T}$ after Debye (cf. p. 96, formula **58**).

$$\frac{F}{T} = \frac{E}{3T} - 3R \log_e \left(1 - e^{-\frac{\beta\nu}{T}}\right).$$

Table II is taken from paper (79), the other tables have been calculated by Frl. Miething.

* For more recent values, see p. 270.   † See note, p. 270.

TABLE I

$\frac{\beta\nu}{T}$ FROM 0 TO 14. C_v AFTER EINSTEIN

$\frac{\beta\nu}{T}$	0·0	0·1	0·2	0·3	0·4	0·5	0·6	0·7	0·8	0·9	1·0
0	5·955	5·947	5·935	5·911	5·878	5·833	5·780	5·718	5·648	5·568	5·483
1	5·483	5·401	5·279	5·184	5·071	4·954	4·832	4·706	4·578	4·446	4·312
2	4·312	4·176	4·039	3·902	3·764	3·626	3·489	3·353	3·218	3·086	2·954
3	2·954	2·827	2·701	2·578	2·458	2·342	2·229	2·119	2·013	1·910	1·811
4	1·811	1·715	1·623	1·536	1·451	1·370	1·292	1·218	1·148	1·081	1·017
5	1·017	0·956	0·898	0·843	0·791	0·743	0·696	0·652	0·610	0·571	0·535
6	0·535	0·499	0·466	0·436	0·407	0·379	0·354	0·330	0·307	0·286	0·266
7	0·266	0·248	0·231	0·215	0·200	0·185	0·172	0·160	0·149	0·138	0·128
8	0·128	0·119	0·110	0·102	0·0945	0·0884	0·0811	0·0752	0·0695	0·0650	0·0600
9	0·0600	0·0554	0·0509	0·0468	0·0435	0·0400	0·0372	0·0340	0·0310	0·0286	0·0266
10	0·0266	0·0247	0·0231	0·0212	0·0196	0·0180	0·0167	0·0152	0·0142	0·0129	0·0119
11	0·0119	0·0110	0·0102	0·0093	0·0085	0·0078	0·0073	0·0067	0·0062	0·0057	0·0052
12	0·0052	0·0048	0·0045	0·0041	0·0038	0·0035	0·0032	0·0029	0·0027	0·0024	0·0022
13	0·0022	0·0020	0·0019	0·0017	0·0016	0·0015	0·0014	0·0013	0·0012	0·0011	0·0010

TABLE II

$\frac{\beta\nu}{T}$ FROM 0 TO 30. C_v AFTER DEBYE

$\frac{\beta\nu}{T}$	0·0	0·1	0·2	0·3	0·4	0·5	0·6	0·7	0·8	0·9	1·0
0	5·955	5·95	5·94	5·93	5·91	5·88	5·85	5·81	5·77	5·72	5·67
1	5·670	5·61	5·55	5·48	5·41	5·34	5·26	5·18	5·09	5·01	4·92
2	4·918	4·83	4·74	4·64	4·54	4·45	4·35	4·25	4·15	4·05	3·95
3	3·948	3·85	3·75	3·65	3·56	3·46	3·36	3·27	3·18	3·09	3·00
4	2·996	2·91	2·82	2·74	2·65	2·57	2·50	2·42	2·34	2·27	2·20
5	2·197	2·13	2·06	1·99	1·93	1·87	1·81	1·75	1·69	1·63	1·58
6	1·582	1·53	1·48	1·43	1·39	1·34	1·30	1·26	1·21	1·18	1·14
7	1·137	1·100	1·065	1·031	0·998	0·966	0·935	0·906	0·878	0·850	0·823
8	0·823	0·798	0·774	0·750	0·727	0·704	0·683	0·662	0·642	0·623	0·604
9	0·604	0·588	0·570	0·552	0·537	0·521	0·507	0·492	0·478	0·465	0·452
10	0·452	0·439	0·427	0·415	0·404	0·394	0·383	0·373	0·363	0·353	0·345
11	0·345	0·335	0·324	0·319	0·310	0·302	0·295	0·287	0·280	0·273	0·267
12	0·267	0·260	0·254	0·248	0·242	0·237	0·231	0·226	0·221	0·216	0·211
13	0·211	0·206	0·202	0·197	0·193	0·188	0·184	0·180	0·176	0·172	0·169
14	0·169	0·165	0·162	0·159	0·155	0·152	0·149	0·146	0·143	0·140	0·137
15	0·137	0·135	0·132	0·130	0·127	0·125	0·122	0·120	0·118	0·116	0·113

$\frac{\beta\nu}{T}$	C_v	$\frac{\beta\nu}{T}$	C_v	$\frac{\beta\nu}{T}$	C_v
16	0·113	21	0·0502	26	0·0264
17	0·0945	22	0·0436	27	0·0236
18	0·0796	23	0·0382	28	0·0212
19	0·0677	24	0·0336	29	0·0190
20	0·0581	25	0·0298	30	0·0172

TABLE IIIA

(a) $\frac{\beta\nu}{T}$ FROM 0 TO 2·00. $\frac{E}{T}$ AFTER EINSTEIN

$\frac{\beta\nu}{T}$	0		1		2		3		4		5		6		7		8		9		10	
0·0	5·955	30	5·925	29	5·896	29	5·867	29	5·838	29	5·809	29	5·780	30	5·750	29	5·721	29	5·792	29	5·663	29
0·1	5·663	29	5·634	29	5·605	29	5·576	28	5·548	28	5·520	28	5·492	28	5·464	28	5·436	28	5·408	28	5·380	28
0·2	5·380	28	5·352	28	5·324	28	5·296	28	5·268	27	5·240	27	5·213	27	5·186	27	5·159	27	5·132	26	5·106	26
0·3	5·106	26	5·080	27	5·053	26	5·027	26	5·001	27	4·974	26	4·948	26	4·921	26	4·895	26	4·869	26	4·843	26
0·4	4·843	26	4·817	26	4·791	25	4·765	25	4·740	25	4·715	25	4·690	25	4·665	25	4·640	25	4·615	25	4·590	25
0·5	4·590	25	4·565	25	4·540	25	4·515	25	4·490	23	4·465	24	4·441	24	4·417	24	4·393	24	4·369	24	4·345	24
0·6	4·345	24	4·321	24	4·297	24	4·273	23	4·250	23	4·227	24	4·203	23	4·180	23	4·157	23	4·134	23	4·111	23
0·7	4·111	23	4·088	23	4·065	22	4·043	22	4·020	22	3·998	22	3·975	22	3·953	22	3·931	22	3·909	22	3·887	22
0·8	3·887	22	3·865	22	3·843	22	3·821	22	3·799	22	3·777	22	3·755	21	3·734	21	3·713	21	3·692	21	3·671	21
0·9	3·671	21	3·650	20	3·630	21	3·609	20	3·589	20	3·569	20	3·558	20	3·527	20	3·507	20	3·486	20	3·466	20
1·0	3·466	21	3·445	20	3·425	20	3·405	20	3·385	20	3·365	20	3·345	20	3·325	19	3·306	19	3·287	19	3·268	19
1·1	3·268	19	3·249	19	3·230	20	3·210	19	3·191	19	3·172	19	3·153	19	3·134	18	3·116	18	3·098	18	3·080	18
1·2	3·080	18	3·062	18	3·044	18	3·026	18	3·008	18	2·990	18	2·972	18	2·954	18	2·936	18	2·918	18	2·900	18
1·3	2·900	18	2·882	17	2·865	17	2·848	17	2·831	17	2·814	17	2·797	17	2·780	17	2·763	17	2·746	17	2·729	17
1·4	2·729	17	2·712	17	2·695	17	2·678	17	2·661	16	2·645	16	2·629	16	2·613	16	2·597	16	2·581	16	2·565	15
1·5	2·565	15	2·550	16	2·534	16	2·518	16	2·502	15	2·487	16	2·471	16	2·455	15	2·440	15	2·425	15	2·410	15
1·6	2·410	15	2·395	15	2·380	15	2·365	15	2·350	15	2·335	15	2·320	15	2·305	14	2·291	14	2·277	14	2·263	14
1·7	2·263	14	2·249	15	2·234	14	2·220	14	2·206	14	2·192	14	2·178	14	2·164	14	2·150	14	2·136	14	2·122	14
1·8	2·122	14	2·108	14	2·094	13	2·081	13	2·068	13	2·055	13	2·042	13	2·029	13	2·016	13	2·003	13	1·990	13
1·9	1·990	13	1·977		1·964	13	1·951	13	1·938	12	1·926	12	1·914	13	1·901	12	1·889	13	1·876	12	1·864	

TABLE IIIb

(b) $\frac{\beta\nu}{T}$ FROM 0 TO 12.0. $\frac{E}{T}$ AFTER EINSTEIN

$\frac{\beta\nu}{T}$	0.0	0.1		0.2		0.3		0.4		0.5		0.6		0.7		0.8		0.9		1.0	
0	5.955	292	5.663	289	5.380	274	5.106	263	4.843	253	4.590	245	4.345	234	4.111	224	3.887	216	3.671	205	3.466
1	3.466	198	3.268	188	3.080	180	2.900	171	2.729	164	2.565	155	2.410	147	2.263	141	2.122	132	1.990	126	1.864
2	1.864	119	1.745	112	1.633	106	1.527	100	1.427	95	1.332	90	1.242	84	1.158	78	1.080	74	1.006	70	0.936
3	0.936	65	0.871	61	0.810	57	0.753	54	0.699	50	0.649	47	0.602	43	0.559	41	0.518	38	0.480	36	0.444
4	0.444	33	0.411	30	0.381	29	0.352	26	0.326	25	0.301	23	0.278	21	0.257	20	0.237	18	0.219	17	0.202
5	0.202	16	0.186	14	0.172	14	0.158	12	0.146	12	0.134	10	0.124	10	0.114	8	0.106	8	0.098	9	0.089
6	0.0890	73	0.0817	66	0.0751	61	0.0690	56	0.0634	51	0.0583	48	0.0535	43	0.0492	40	0.0452	38	0.0415	35	0.0380
7	0.0380	31	0.0349	29	0.0320	26	0.0294	25	0.0269	21	0.0248	21	0.0227	19	0.0208	18	0.0190	16	0.0174	14	0.0160
8	0.0160	14	0.0146	12	0.0134	11	0.0123	10	0.0113	10	0.0103	9	0.0094	8	0.0086	7	0.0079	6	0.0072	6	0.0066
9	0.0066	5	0.0061	6	0.0055	5	0.0050	4	0.0046	4	0.0042	3	0.0039	4	0.0035	3	0.0032	2	0.0030	3	0.0027
10	0.0027	2	0.0025	2	0.0023	2	0.0021	2	0.0019	2	0.0017	1	0.0016	2	0.0014	1	0.0013	1	0.0012	1	0.0011
11	0.0011	1	0.0010	1	0.0009	1	0.0008	0	0.0008	1	0.0007	1	0.0006	0	0.0006	1	0.0005	0	0.0005	1	0.0004
12	0.0004																				

TABLE IVA

(a) $\dfrac{\beta\nu}{T}$ FROM 0 TO 2·00. $\dfrac{F}{T}$ AFTER EINSTEIN

$\dfrac{\beta\nu}{T}$	0		1		2		3		4		5		6		7		8		9		10
0·1	14·01	57	13·44	51	12·93	46	12·47	40	12·07	37	11·68	36	11·32	33	10·99	31	10·68	26	10·42	25	10·17
0·2	10·17	25	9·92	24	9·68	24	9·44	23	9·21	23	8·98	21	8·77	20	8·57	19	8·38	17	8·21	17	8·04
0·3	8·04	17	7·87	16	7·71	16	7·55	15	7·40	14	7·26	13	7·13	14	6·99	13	6·86	12	6·74	13	6·61
0·4	6·61	11	6·50	11	6·39	11	6·28	11	6·17	10	6·07	11	5·96	10	5·86	10	5·76	10	5·66	10	5·56
0·5	5·56	9	5·47	9	5·38	10	5·28	9	5·19	9	5·10	8	5·02	7	4·95	7	4·88	7	4·81	7	4·74
0·6	4·74	8	4·66	7	4·59	7	4·52	7	4·45	6	4·39	7	4·32	6	4·26	6	4·20	6	4·14	6	4·08
0·7	4·08	6	4·02	5	3·97	5	3·92	6	3·86	5	3·81	5	3·76	5	3·71	5	3·66	5	3·61	6	3·55
0·8	3·55	4	3·51	4	3·47	5	3·42	4	3·38	4	3·34	5	3·29	4	3·25	4	3·21	5	3·16	5	3·11
0·9	3·11	4	3·07	4	3·03	3	3·00	4	2·96	3	2·93	5	2·88	3	2·85	4	2·81	4	2·77	4	2·73
1·0	2·73	39	2·692	37	2·655	35	2·620	33	2·587	32	2·556	32	2·523	29	2·494	28	2·466	28	2·438	29	2·409
1·1	2·409	29	2·380	28	2·352	28	2·324	27	2·297	27	2·270	27	2·243	28	2·215	27	2·188	27	2·161	27	2·134
1·2	2·134	27	2·107	26	2·081	25	2·056	24	2·032	24	2·008	24	1·984	23	1·961	23	1·938	22	1·916	22	1·894
1·3	1·894	22	1·872	22	1·850	22	1·828	21	1·807	21	1·786	20	1·766	20	1·746	20	1·726	20	1·706	20	1·686
1·4	1·686	19	1·667	19	1·648	19	1·629	16	1·613	16	1·597	19	1·572	18	1·554	18	1·536	17	1·519	17	1·502
1·5	1·502	16	1·486	16	1·470	16	1·454	16	1·438	15	1·423	17	1·406	16	1·390	16	1·374	16	1·358	16	1·342
1·6	1·342	15	1·327	15	1·312	15	1·297	14	1·283	14	1·269	14	1·255	14	1·241	14	1·227	13	1·214	14	1·200
1·7	1·200	13	1·187	14	1·173	13	1·160	12	1·148	13	1·135	12	1·123	12	1·111	12	1·099	12	1·087	12	1·075
1·8	1·075	12	1·063	12	1·051	11	1·040	11	1·029	12	1·017	12	1·005	11	0·994	11	0·983	11	0·972	12	0·964
1·9	0·964	12	0·952	11	0·941	11	0·930	10	0·920	10	0·910	10	0·900	9	0·891	9	0·882	8	0·873	9	0·864

TABLE IVb

(b) $\frac{\beta\nu}{T}$ FROM 0 TO 7·0. $\frac{F}{T}$ AFTER EINSTEIN

$\frac{\beta\nu}{T}$	0·0		0·1		0·2		0·3		0·4		0·5		0·6		0·7		0·8		0·9		1·0
0	∞	—	14,006	3836	10·170	2129	8·041	1333	6·608	1048	5·560	820	4·740	659	4·081	535	3·546	440	3·106	375	2·731
1	2·731	322	2·409	275	2·134	240	1·894	208	1·686	184	1·502	160	1·342	142	1·200	125	1·075	111	0·964	100	0·864
2	0·864	88	0·776	78	0·698	70	0·628	63	0·565	56	0·509	51	0·458	46	0·412	40	0·372	37	0·335	33	0·302
3	0·302	29	0·273	27	0·246	24	0·222	22	0·200	19	0·181	17	0·164	16	0·148	15	0·133	14	0·119	12	0·107
4	0·107	10	0·097	9	0·088	9	0·079	8	0·071	7	0·064	6	0·058	6	0·052	5	0·047	5	0·042	4	0·038
5	0·038	4	0·034	3	0·031	3	0·028	3	0·025	3	0·022	2	0·020	2	0·018	2	0·016	2	0·014	1	0·013
6	0·013	2	0·011	2	0·009	1	0·008	1	0·007	1	0·006	1	0·005	1	0·004	0	0·004	1	0·003	0	0·003
7	0·003																				

TABLE Vᴀ

(a) $\frac{\beta\nu}{T}$ FROM 0 TO 2.00. $\frac{E}{T}$ AFTER DEBYE

$\frac{\beta\nu}{T}$	0		1		2		3		4		5		6		7		8		9		10
0.1	5.733	23	5.710	22	5.688	21	5.667	21	5.646	21	5.625	21	5.604	21	5.583	21	5.562	21	5.541	21	5.520
0.2	5.520	20	5.500	20	5.480	21	5.459	21	5.438	21	5.417	21	5.396	21	5.375	21	5.354	21	5.333	21	5.312
0.3	5.312	21	5.291	20	5.271	21	5.250	20	5.230	20	5.210	20	5.190	20	5.170	20	5.150	20	5.130	20	5.110
0.4	5.110	19	5.091	20	5.071	20	5.051	20	5.031	19	5.012	20	4.992	20	4.972	20	4.952	19	4.933	20	4.913
0.5	4.913	20	4.893	19	4.874	19	4.855	19	4.836	19	4.817	19	4.788	19	4.779	19	4.760	19	4.741	19	4.722
0.6	4.722	18	4.704	19	4.685	19	4.666	19	4.647	19	4.628	18	4.610	18	4.592	18	4.574	19	4.555	19	4.536
0.7	4.536	18	4.518	18	4.500	17	4.483	18	4.465	18	4.447	18	4.429	17	4.412	18	4.394	18	4.376	18	4.358
0.8	4.358	17	4.341	17	4.324	17	4.307	17	4.290	17	4.273	18	4.255	17	4.238	17	4.221	18	4.203	17	4.186
0.9	4.186	17	4.169	17	4.152	17	4.135	17	4.118	17	4.101	17	4.084	17	4.067	17	4.050	17	4.033	16	4.017
1.0	4.017	16	4.001	16	3.985	17	3.968	16	3.952	17	3.935	17	3.918	16	3.902	16	3.886	16	3.870	16	3.854
1.1	3.854	16	3.838	16	3.822	16	3.806	16	3.790	16	3.774	16	3.758	16	3.742	16	3.726	16	3.710	15	3.695
1.2	3.695	15	3.680	15	3.665	15	3.650	15	3.635	15	3.620	15	3.605	15	3.590	15	3.575	15	3.560	15	3.545
1.3	3.545	15	3.530	15	3.515	15	3.500	14	3.486	15	3.471	14	3.457	15	3.442	14	3.428	15	3.413	14	3.399
1.4	3.399	14	3.385	14	3.371	14	3.357	14	3.343	14	3.329	14	3.315	14	3.301	14	3.287	14	3.273	14	3.259
1.5	3.259	14	3.245	14	3.231	14	3.217	14	3.203	13	3.190	14	3.176	13	3.163	13	3.150	14	3.136	13	3.123
1.6	3.123	13	3.110	14	3.096	14	3.082	13	3.069	13	3.056	13	3.043	13	3.030	13	3.017	13	3.004	12	2.992
1.7	2.992	13	2.979	13	2.966	13	2.953	13	2.940	13	2.927	12	2.915	13	2.902	12	2.890	13	2.877	13	2.864
1.8	2.864	13	2.851	12	2.839	13	2.826	12	2.814	13	2.801	12	2.789	13	2.776	12	2.764	12	2.752	13	2.739
1.9	2.739	12	2.727	11	2.716	12	2.704	12	2.692	11	2.681	11	2.670	11	2.659	11	2.648	11	2.637	10	2.627

TABLE V$_B$

(b) $\frac{\beta\nu}{T}$ FROM 0 TO 16·0. $\frac{E}{T}$ AFTER DEBYE

$\frac{\beta\nu}{T}$	0·0	0·1	0·2	0·3	0·4	0·5	0·6	0·7	0·8	0·9	1·0
0	5·955	—	5·5195	5·3122	5·1100	4·9130	4·7220	4·5364	4·3578	4·1862	4·0168
1	4·0168	3·8535	3·6951	3·5450	3·3991	3·2592	3·1229	2·9920	2·8640	2·7395	2·6266
2	2·6266	2·5138	2·4068	2·3047	2·2044	2·1078	2·0166	1·9288	1·8446	1·7642	1·6873
3	1·6873	1·6131	1·5423	1·4756	1·4118	1·3492	1·2917	1·2364	1·1825	1·1314	1·0821
4	1·0821	1·0361	0·9931	0·9517	0·9118	0·8733	0·8361	0·8002	0·7654	0·7317	0·7009
5	0·7009	0·6712	0·6438	0·6187	0·5944	0·5708	0·5478	0·5255	0·5037	0·4824	0·4618
6	0·4618	0·4437	0·4259	0·4088	0·3926	0·3787	0·3652	0·3519	0·3387	0·3257	0·3128
7	0·3128	0·3017	0·2908	0·2803	0·2702	0·2605	0·2513	0·2423	0·2340	0·2263	0·2195
8	0·2195	0·2135	0·2077	0·2017	0·1959	0·1905	0·1855	0·1797	0·1744	0·1691	0·1639
9	0·1639	0·1588	0·1536	0·1485	0·1435	0·1384	0·1336	0·1289	0·1242	0·1195	0·1149
10	0·1149	0·1107	0·1070	0·1038	0·1009	0·0983	0·0957	0·0931	0·0907	0·0886	0·0866
11	0·0866	0·0845	0·0824	0·0804	0·0783	0·0763	0·0742	0·0722	0·0704	0·0686	0·0671
12	0·0671	0·0655	0·0640	0·0625	0·0610	0·0595	0·0580	0·0565	0·0552	0·0540	0·0526
13	0·0526	0·0514	0·0502	0·0491	0·0481	0·0471	0·0461	0·0451	0·0441	0·0431	0·0420
14	0·0420	0·0411	0·0403	0·0395	0·0388	0·0380	0·0373	0·0365	0·0358	0·0350	0·0343
15	0·0343	0·0335	0·0328	0·0320	0·0313	0·0308	0·0303	0·0298	0·0293	0·0288	0·0283

TABLE VIa

$(a)\ \dfrac{\beta\nu}{T}$ FROM 0 TO 2·00. $\dfrac{F}{T}$ AFTER DEBYE

$\dfrac{\beta\nu}{T}$	0		1		2		3		4		5		6		7		8		9		10
0·1	15·92	47	15·45	45	15·00	44	14·56	42	14·14	40	13·74	35	13·39	35	13·04	34	12·70	35	12·35	34	12·01
0·2	12·01	28	11·73	23	11·50	23	11·27	24	11·03	23	10·80	23	10·57	21	10·36	19	10·17	19	9·98	17	9·81
0·3	9·81	17	9·64	15	9·49	15	9·34	15	9·19	15	9·04	15	8·89	15	8·74	15	8·59	14	8·45	14	8·31
0·4	8·31	13	8·18	12	8·06	11	7·95	12	7·83	11	7·72	11	7·61	11	7·50	10	7·40	11	7·29	10	7·19
0·5	7·19	10	7·09	10	6·99	10	6·89	9	6·80	9	6·71	9	6·62	8	6·54	8	6·46	8	6·38	7	6·31
0·6	6·31	8	6·23	8	6·15	7	6·08	8	6·00	7	5·93	7	5·86	7	5·79	7	5·72	7	5·65	5	5·60
0·7	5·60	8	5·52	6	5·46	6	5·40	6	5·34	6	5·28	5	5·23	6	5·17	5	5·12	5	5·07	6	5·01
0·8	5·006	51	4·950	50	4·905	50	4·855	51	4·804	50	4·754	50	4·704	50	4·654	50	4·604	51	4·553	50	4·503
0·9	4·503	46	4·461	45	4·412	46	4·360	44	4·322	44	4·278	42	4·236	40	4·196	40	4·156	39	4·117	40	4·077
1·0	4·077	39	4·038	39	3·999	39	3·960	39	3·921	38	3·883	39	3·844	39	3·805	38	3·767	36	3·731	36	3·695
1·1	3·695	35	3·660	35	3·625	35	3·590	34	3·556	33	3·528	32	3·491	32	3·459	32	3·427	32	3·496	30	3·365
1·2	3·365	30	3·335	30	3·305	28	3·275	30	3·245	30	3·215	29	3·186	28	3·158	27	3·131	28	3·103	27	3·076
1·3	3·076	27	3·049	27	3·022	26	2·996	27	2·969	27	2·942	26	2·916	25	2·891	24	2·867	24	2·843	24	2·819
1·4	2·819	23	2·796	23	2·773	23	2·750	24	2·726	23	2·703	23	2·680	23	2·657	23	2·634	22	2·612	22	2·590
1·5	2·590	22	2·568	21	2·547	21	2·526	20	2·506	21	2·485	21	2·464	20	2·444	20	2·424	20	2·404	20	2·384
1·6	2·384	19	2·365	19	2·346	18	2·328	18	2·310	19	2·291	18	2·273	18	2·255	19	2·236	18	2·218	19	2·199
1·7	2·199	18	2·181	17	2·164	17	2·147	17	2·130	16	2·114	17	2·097	17	2·080	17	2·063	16	2·047	16	2·031
1·8	2·031	15	2·016	15	2·001	16	1·985	16	1·969	15	1·954	15	1·939	15	1·924	16	1·908	15	1·893	15	1·878
1·9	1·878	15	1·863	14	1·849	14	1·835	14	1·821	14	1·807	14	1·793	14	1·779	13	1·766	13	1·753	12	1·741

TABLE VIb

(b) $\frac{\beta\nu}{T}$ FROM 0 TO 15·0. $\frac{F}{T}$ AFTER DEBYE

$\frac{\beta\nu}{T}$	0·0	0·1	0·2	0·3	0·4	0·5	0·6	0·7	0·8	0·9	1·0
0	—	15·9180	12·0098	9·8111	8·3113	7·1921	6·3134	5·5993	5·0065	4·5030	4·0766
1	4·0766	3·6949	3·3650	3·0756	2·8192	2·5899	2·3839	2·1986	2·0307	1·8781	1·7414
2	1·7414	1·6158	1·5016	1·3973	1·3011	1·2125	1·1318	1·0573	0·9886	0·9251	0·8665
3	0·8665	0·8122	0·7619	0·7157	0·6731	0·6324	0·5954	0·5612	0·5290	0·4992	0·4708
4	0·4708	0·4449	0·4210	0·3985	0·3774	0·3576	0·3389	0·3212	0·3044	0·2885	0·2739
5	0·2739	0·2605	0·2476	0·2361	0·2251	0·2146	0·2047	0·1951	0·1860	0·1771	0·1688
6	0·1688	0·1613	0·1540	0·1474	0·1408	0·1351	0·1298	0·1246	0·1196	0·1146	0·1097
7	0·1097	0·1055	0·1014	0·0974	0·0937	0·0901	0·0868	0·0835	0·0804	0·0776	0·0751
8	0·0751	0·0730	0·0709	0·0687	0·0667	0·0646	0·0629	0·0609	0·0590	0·0572	0·0554
9	0·0554	0·0536	0·0518	0·0500	0·0483	0·0466	0·0449	0·0433	0·0417	0·0401	0·0386
10	0·0386	0·0371	0·0358	0·0348	0·0338	0·0329	0·0320	0·0311	0·0303	0·0296	0·0289
11	0·0289	0·0282	0·0275	0·0268	0·0261	0·0254	0·0247	0·0241	0·0235	0·0229	0·0224
12	0·0224	0·0218	0·0213	0·0208	0·0203	0·0198	0·0193	0·0188	0·0184	0·0180	0·0175
13	0·0175	0·0172	0·0167	0·0164	0·0160	0·0157	0·0154	0·0150	0·0147	0·0144	0·0140
14	0·0140	0·0137	0·0134	0·0132	0·0129	0·0126	0·0124	0·0122	0·0119	0·0117	0·0114

LIST OF THERMODYNAMICAL PAPERS FROM THE PHYSICO-CHEMICAL INSTITUTE OF THE UNIVERSITY OF BERLIN (1906-1916)

1906

1. W. NERNST, Über die Berechnung chemischer Gleichgewichte aus thermischen Messungen. Kgl. Ges. d. Wiss. Gött., Heft I.
2. L. LÖWENSTEIN, Eine neue Methode zur Untersuchung von Gasgleichgewichten bei hoher Temperatur. Zeitschr. f. physikal. Chem., **54**, S. 707.
3. W. NERNST und H. v. WARTENBERG, I. Die Dissoziation von Wasserdampf. II. Über die Dissoziation der Kohlensäure. Zeitschr. f. physikal. Chem., **56**, S. 513.
4. O. BRILL, Über die Dampfspannungen von flüssigem Ammoniak. Ann. d. Physik [4], **21**, S. 170.
5. W. NERNST, Über die Beziehungen zwischen Wärmeentwicklung und maximaler Arbeit bei kondensierten Systemen. Sitzungsber. d. Kgl. Pr. Akad. d. Wiss. vom 20 Dez.
5A. E. C. BINGHAM, Vapor-Pressure and Chemical Composition. Journal of American Chem. Society, **28**, S. 717.

1907

6. H. v. WARTENBERG, Über das Cyan, Cyanwasserstoff-und Acetylengleichgewicht. Zeitschr. f. anorg. Chem., **52**, S. 299.
7. O. BRILL, Zur Berechnung der Dampfdichten dissoziierender Substanzen. Zeitschr. f. physikal. Chem., **57**, S. 721.
8. W. NERNST, Über das Ammoniakgleichgewicht. Zeitschr. f. Elektrochem., **13**, S. 521.
9. G. PREUNER, Dissoziation des Schwefelwasserstoffs. Zeitschr. f. anorg. Chem., **55**, S. 279.
10. H. v. WARTENBERG, Zur Berechnung von Kohlenwasserstoffgleichgewichten. Zeitschr. f. physikal. Chem., **61**, S. 366.
11. R. NAUMANN, Experimentelle Bestimmung und theoretische Berechnung kleiner Dampfdrucke von Jod und Metalljodiden. Inaug.-Diss. Universität Berlin. 21 Dez. 1907.

1908

12. M. BODENSTEIN und G. DUNANT, Die Dissoziation des Kohlenoxychlorids. Zeitschr. f. physikal. Chem., **61**, S. 437.

13. F. JOST, 1. Über die Einwirkung von Wasserstoff auf Kohle bei hohen Temperaturen. 2. Über das Ammoniakgleichgewicht. Inaug.-Diss. Univ. Berlin. 15 Apr., 1908. Zeitschr. f. anorg. Chem., **57**, S. 414.

14. F. HALLA, Zur thermodynamischen Berechnung elektromotorischer Kräfte. Zeitschr. f. Elektrochem., **14**, S. 411.

15. F. JOST, Über die Lage des Ammoniakgleichgewichtes. Zeitschr. f. Elektrochem., **14**, S. 373.

16. H. SCHOTTKY, Studien zur Thermodynamik der kristallwasserhaltigen Salze. Zeitschr. f. physikal. Chem., **64**, S. 415.

17. E. FALCK, Theoretische Bestimmung des Dampfdrucks fester und flüssiger Kohlensäure. Physikal. Zeitschr., **9**, S. 433.

18. F. VOLLER, Über eine neue Methode zur direkten Bestimmung der spezifischen Wärme der Gase bei konstantem Volumen. Inaug.-Diss. Univ. Berlin. 9 Okt., 1908.

1909

19. W. NERNST, Über die Berechnung elektromotorischer Kräfte aus thermischen Messungen. Ber. d. Kgl. Pr. Akad. d. Wiss. vom 11 Feb.

20. H. SCHLESINGER, Die spezifischen Wärmen von Lösungen I. Physikal. Zeitschr., **10**, S. 210.

21. M. WASJUCHNOWA, Das Gleichgewicht Kupri-Kuprosulfid. Inaug.-Diss. Univ. Berlin. 9 Okt.

22. H. SCHOTTKY, Messungen von spezifischen Wärmen mit einem neuen Flüssigkeitskalorimeter. Physikal. Zeitschr., **10**, S. 634.

23. A. EUCKEN, Über die Bestimmung, spezifischer Wärmen bei tiefen Temperaturen. Physikal. Zeitschr., **10**, S. 586.

24. M. PIER, Die spezifischen Wärmen von Argon, Wasserdampf, Stickstoff, Wasserstoff bei sehr hohen Temperaturen. Zeitschr. f. Elektrochem., **15**, S. 536.

25. W. NERNST, Die chemische Konstante des Wasserstoffs und seine Affinität zu den Halogenen. Zeitschr. f. Elektrochem., **15**, S. 687.

26. F. POLLITZER, Über das Gleichgewicht der Reaktion $H_2S + 2J = 2HJ + S$ und die Dissoziation von H_2S. Zeitschr. f. anorg. Chem., **64**, S. 121.

27. H. Levy, Thermodynamische Behandlung einiger Eigenschaften des Wassers II. Verh. d. Deutsch. Physikal. Ges., **11**, S. 328.

28. W. Nernst, Thermodynamische Behandlung einiger Eigenschaften des Wassers I. und III. Verh. d. Deutsch. Physikal. Ges., **11**, S. 313, 336.

29. F. Horak, Die Dissoziation des Kohlenoxychlorids. Inaug.-Diss. Univ. Berlin. 14 Okt.

29A. H. v. Wartenberg, Bildungswärmen von Kupro- und Kupri-sulfid. Zeitschr. physikal. Chem., **67**, S. 446.

29B. M. Bodenstein und Katayama, Dissoziation von hydratischer Schwefelsäure und von Stickstoffdioxyd. Zeitschr. f. Elektrochem., **15**, S. 244.

1910

30. J. Barker, Experimentelle Bestimmung und thermodynamische Berechnung der Dampfdrucke von Toluol, Naphtalin und Benzol. Zeitschr. f. physikal. Chem., **71**, S. 235.

31. F. Koref, Über das Gleichgewicht bei Schwefelkohlenstoffbildung. Zeitschr. f. anorg. Chem., **66**, S. 73.

32. H. Levy, Thermodynamische Behandlung einiger Eigenschaften des Wassers und des Wasserdampfes. Inaug.-Diss. Univ. Berlin.

33. W. Nernst, Spezifische Wärme und chemisches Gleichgewicht des Ammoniakgases. Zeitschr. f. Elektrochem., **16**, S. 96.

34. A. Magnus, Über die Bestimmung spezifischer Wärmen. Ann. d. Phys., **31**, S. 597.

35. W. Nernst, Sur les chaleurs spécifiques aux basses températures et le développement de la thermodynamique. Séances de la soc. franc. de phys., 1er fascicule.

36. W. Nernst, F. Koref, und F. A. Lindemann, Untersuchungen über die spezifische Wärme bei tiefen Temperaturen I und II. Ber. d. Kgl. Pr. Akad. d. Wiss. vom 3 März.

37. A. Magnus und F. A. Lindemann, Über die Abhängigkeit der spezifischen. Wärme fester Körper von der Temperatur. Zeitschr. f. Elektrochem., **16**, S. 269.

38. W. Nernst, Sur la détermination de l'affinité chimique à partir des données thermiques. Journ. de chimie phys., **8**, S. 225.

39. W. Nernst, Thermodynamische Berechnung des Dampfdruckes von Wasser und Eis. Verh. d. Deutsch. Physikal. Ges., **12**, S. 565.

LIST OF LITERATURE

40. F. A. LINDEMANN, Über die Berechnung molekularer Eigenfrequenzen. Physikal. Zeitschr., **14**, S. 609.
41. W. NERNST, The specific heat of ice, water, and water vapour. Transactions of the Faraday Soc., vol. vi, part 1.
42. W. NERNST, Neuere Entwicklung der Theorie der galvanischen Elemente. Zeitschr. f. Elektrochem., **16**, S. 517.
43. M. PIER, Spezifische Wärmen und Gasgleichgewiche nach Explosionsversuchen II. Zeitschr. f. Elektrochem., **16**, S. 897.
44. F. KEUTEL, Über die spezifische Wärme von Gasen. Inaug.-Diss. Univ. Berlin. 1 Aug.

1911

45. C. HOLLAND, Untersuchungen mit dem Quarzmanometer, die Dissoziation der gasförmigen Essigsäure und des Phosphorpentachlorids. Inaug.-Diss. Univ. Berlin. 12 Aug.
46. F. POLLITZER, Bestimmungen spezifischer Wärmen bei tiefen Temperaturen und ihre Verwertung zur Berechnung elektromotorischer Kräfte. Zeitschr. f. Elektrochem., I, **17**, S. 5 ; II, **19**, S. 515.
47. W. NERNST, Der Energieinhalt fester Stoffe. Ann. d. Physik, **36**, S. 395.
48. W. NERNST, Über neuere Probleme der Wärmetheorie. Ber. d. Kgl. Pr. Akad. d. Wiss. vom 26 Jan.
49. W. NERNST, Untersuchungen über die spezifische Wärme bei tiefen Temperaturen III. Ber. d. Kgl. Pr. Akad. d. Wiss. vom 9 März.
50. F. HALLA, Zur thermodynamischen Berechnung elektromotorischer Kräfte II. Zeitschr. f. Elektrochem., **17**, S. 179.
51. W. NERNST, Zur Theorie der spezifischen Wärme und über die Anwendung der Lehre von den Energiequanten auf physikalischchemische Fragen überhaupt. Zeitschr. f. Elektrochem, **17**, S. 265.
52. W. NERNST und F. A. LINDEMANN, Untersuchungen über die spezifische Wärme bei tiefen Temperaturen V. Ber. d. Kgl. Pr. Akad. d. Wiss. v. 27 April.
53. F. LINDEMANN, Untersuchung über die spezifische Wärme bei tiefen Temperaturen IV. Ber. d. Kgl. Pr. Akad. d. Wiss. vom 9 März.
54. R. THIBAUT, Die spezifische Wärme verschiedener Gase und Dämpfe. Ann. d. Physik, **35**, S. 347.
55. C. L. LINDEMANN, Über die Temperaturabhängigkeit des thermischen Ausdehnungskoeffizienten. Physikal. Zeitschr., **12**, S. 1197.
56. N. BJERRUM, Über die spezifische Wärme der Gase. Zeitschr. f. Elektrochem., **17**, S. 731.

57. W. Nernst, Über einen Apparat zur Verflüssigung von Wasserstoff. Zeitschr. f. Elektrochem., **17**, S. 735.

58. C. L. Lindemann, Über die Temperaturabhängigkeit des thermischen Ausdehnungskoeffizienten. Physikal. Zeitschr., **12**, S. 1197.

59. F. A. Lindemann, Über das Dulong-Petitsche Gesetz. Inaug.-Diss. Univ. Berlin. 17 Juli.

60. F. Koref, Messungen der spezifischen Wärmen bei tiefen Temperaturen mit dem Kupferkalorimeter. Ann. d. Physik, **36**, S. 49.

61. W. Nernst und F. A. Lindemann, Spezifische Wärme und Quantentheorie. Zeitschr. f. Elektrochem., **17**, S. 817.

62. H. Levy, Über die Molekularwärme des Wasserdampfes. Verh. d. Deutsch. Physikal.-Ges., **13**, S. 926.

62A. W. Nernst, Über ein allgemeines Gesetz, das Verhalten fester Stoffe bei sehr tiefen Temperaturen betreffend. Verh. d. Deutsch. Physik. Ges., **13**, S. 921.

63. F. Pollitzer, Zur Thermodynamik des Clarkelements. Zeitschr. f. physikal. Chem., **78**, S. 374.

1912

64. A. Eucken, Die Molekularwärme des Wasserstoffs bei tiefen Temperaturen. Ber. d. Kgl. Pr. Akad: d. Wiss. vom 1 Febr.

65. W. Nernst, Thermodynamik und spezifische Wärme. Ber. d. Kgl. Pr. Akad. d. Wiss. vom 1 Febr.

66. W. Nernst, Energieinhalt der Gase. Physikal. Zeitschr., **13**, S. 1064.

67. H. Budde, Die Dissoziation des Schwefeldampfes in die Atome. Zeitschr. f. anorg. Chem., **78**, S. 169.

68. W. Nernst, Zur neueren Entwicklung der Thermodynamik. Ges. deutscher Naturf. und Ärzte. Verh. 1912.

69. H. v. Wartenberg, Über die Reduktion der Kieselsäure. Zeitschr. f. anorg. Chem., **19**, S. 71.

70. C. Holland, Die Dissoziation der gasförmigen Essigsäure und des Phosphorpentachlorids. Zeitschr. f. Elektrochem., **18**, S. 234; **22**, S. 37.

71. F. A. Lindemann, Some considerations on the forces acting between the atoms of solid bodies. Nernst-Festschrift bei Knapp (Halle).

72. N. Bjerrum (1) Die Dissoziation und die spezifische Wärme von Wasserdampf bei sehr hohen Temperaturen nach Explosionsversuchen; (2) Die Dissoziation und die spezifische Wärme von CO_2

bei sehr hohen Temperaturen nach Explosionsversuchen. Zeitschr. f. physikal. Chem., **79**, S. 513.

73. N. BJERRUM, Über das Verhalten von Jod und Schwefel bei extrem hohen Temperaturen nach Explosionsversuchen. Zeitschr. f. physikal. Chem., **79**, S. 281.

74. U. FISCHER, Über die Affinität zwischen Jod und Silber. Zeitschr. f. anorg. Chem., **78**, S. 41.

75. H. BUDDE, Die Molekularwärme des Ammoniaks. Zeitschr. f. anorg. Chem., **78**, S. 159.

76. F. KOREF, Über die Eigenfrequenzen von Elementen in Verbindungen. Physikal. Zeitschr., **13**, S. 183.

77. F. KOREF und H. BRAUNE, Die Bildungswärme von Bleijodid und Bleichlorid. Zeitschr. f. Elektrochem., **18**, S. 818.

78. W. NERNST und F. A. LINDEMANN : (1) Untersuchungen über die spezifische Wärme VI ; (2) Berechnung von Atomwärmen.

79. W. NERNST, Zur Berechnung chemischer Affinitäten VII. Ber. d. Kgl. Pr. Akad. d. Wiss. vom 12 Dez.

79A. A. S. RUSSELL, Messungen von spezifischen Wärmen bei tiefen Temperaturen. Physikal. Zeitschr., **13**, S. 59.

1913

80. A. EUCKEN, Über die Berechnung spezifischer Wärmen aus elastischen Konstanten. Verh. d. Deutsch. Physikal. Ges., **15**, S. 571.

81. A. SIGGEL, Thermodynamische Untersuchungen am Kupfersulfat. Zeitschr. f. Elektrochem., **19**, S. 340.

82. A. EUCKEN, Über das Wärmeleitungsvermögen, die spezifische Wärme und innere Reibung der Gase. Physikal. Zeitschr., **14**, S. 324.

83. W. NERNST, Das Gleichgewichtsdiagramm der beiden Schwefelmodifikationen. Zeitschr. f. physikal. Chem., **83**, S. 546.

84. A. EUCKEN und F. SCHWERS, Eine experimentelle Prüfung des T^3-Gesetzes für den Verlauf der spezifischen Wärme fester Körper bei tiefen Temperaturen. Verh. d. Deutsch. Physikal. Ges., **15**, S. 578.

85. F. BULLE, Über die Dampfdruckkurve des Sauerstoffs und über eine Bestimmung der kritischen Daten von Wasserstoff. Phys. Zeitschr., **14**, S. 860.

86. G. WIETZEL, Über das thermoelektrische Verhalten der Metalle bei tiefen Temperaturen. Ann. d. Physik (4), **43**, S. 605.

87. F. GERMANN, Eine Bestimmung der Dampfdruck- und Dichtekurven des Sauerstoffs und Konstruktion eines Apparates zur Bestimmung kritischer Daten. Phys. Zeitschr., **14**, S. 857.

88. J. R. PARTINGTON, Eine Bestimmung des Verhältnisses der spezifischen Wärme der Luft und Kohlensäure sowie eine Berechnung der spezifischen Wärme mittels der Berthelotschen Zustandsgleichung. Phys. Zeitschr., **14**, S. 969.

89. W. NERNST, Über den maximalen Nutzeffekt von Verbrennungsmotoren. Zeitschr. f. Elektrochem., **19**, S. 699.

90. L. WOLFF, Über die Ermittlung von Bildungswärmen mit Hilfe elektromotorischer Kräfte. Inaug.-Diss. Univ. Berlin, 14 Okt. ; Zeitschr. f. Elektrochem., **20**, S. 19.

91. H. v. SIEMENS, Über Dampfdruckmessungen und Thermometrie bei tiefen Temperaturen. Ann. d. Physik, **42**, S. 871.

92. C. F. MÜNDEL, Experimentelle Bestimmung und theoretische Berechnung kleiner Dampfdrucke bei tiefen Temperaturen. Z. f. physikal. Chem., **85**, S. 435.

93. W. NERNST, Zur Thermodynamik kondensierter Systeme. Ber. d. Kgl. Pr. Akad. d. Wiss. vom 11 Dez.

1914

94. H. SCHIMANK, Über das Verhalten des elektrischen Widerstandes von Metallen bei tiefen Temperaturen. Ann. d. Physik, **45**, S. 706.

95. W. NERNST and F. SCHWERS, Untersuchungen über die spezifische Wärme bei tiefen Temperaturen VIII. Ber. d. Kgl. Pr. Akad. d. Wiss.

95A. W. NERNST, Anwendnung des Wärmesatzes auf Gase. Zeitschr. f. Elektrochem., **20**, S. 357.

96. P. WINTERNITZ, Über die Einwirkung von Schwefel auf CO und CO_2 und die SO_2-Dissoziation. Inaug.-Diss. Univ. Berlin, 4 März.

97. W. SIEGEL, Untersuchungen von Gasgleichgewichten und spezifischen Wärmen nach der Explosionsmethode. Z. f. phys. Chem., **87**, S. 641.

98. v. KOHNER und P. WINTERNITZ, Die chemische Konstante des Wasserstoffs. Phys. Zeitschr., **15**, S. 393 u. 645.

98A. P. WINTERNITZ, Über eine Anwendung der Nernstschen Näherungsformel. Phys. Zeitschr., **15**, S. 397.

98B. BRAUNE u. KOREF, Messung der Bildungswärmen von Blei- und Silberhalogeniden und ihre Anwendung zur Prüfung des Nernstschen Wärmetheorems. Zeitschr. f. anorg. Chem., **87**, S. 175.

99. R. EWALD, Messungen spezifischer Wärmen und Beiträge zur Molekulargewichtsbestimmung. Ann. d. Physik, **44**, S. 1213.

LIST OF LITERATURE

100. K. v. Kohner, Über die Dissoziationsspannung des Kalziumkarbonates. Inaug.-Diss. Univ. Berlin. 5 Aug.
101. A. Eucken, Über den Quanteneffekt bei einatomigen Gasen und Flüssigkeiten. Ber. d. Kgl. Pr. Akad. d. Wiss.
102. W. Drägert, Thermodynamische Untersuchungen am Kalziumhydroxyd sowie über die graphische und mechanische Berechnung chemischer Affinitäten aus thermischen Messungen. Inaug.-Diss. Univ. Berlin. 5 Aug.

1915

102A. T. Isnardi, Wärmeleitung in dissoziierten Gasen und Dissoziation des Wasserstoffs in Atome. Zeitschr. f. Elektrochem., **21**, S. 405.

1916

103. W. Nernst, Über die experimentelle Bestimmung chemischer Konstanten. Zeitscher. f. Elektrochem., **22**, S. 185.
104. H. Schimank, Über die Anwendbarkeit der Daniel-Berthelotschen Zustandsgleichung auf das Verhalten von Dämpfen. Physik. Zeitschr., **17**, S. 393.
105. A. Eucken, Über das thermische Verhalten einiger komprimierter und kondensierter Gase bei tiefen Temperaturen. Verh. d. Deutsch. Physikal. Ges., **18**, S. 4.
106. A. Eucken, Über den Wärmeinhalt einatomiger Flüssigkeiten. Verh. d. Deutsch. Physikal. Ges., **18**, S. 18.
107. P. Günther, Untersuchungen über die spezifische Wärme bei tiefen Temperaturen. Ann. d. Physik, **51**, S. 828.
108. W. Nernst, Über einen Versuch, von quantentheoretischen Betrachtungen zur Annahme stetiger Energieänderungen zurückzukehren. Verh. d. Deutsch. Physikal. Ges., **18**, S. 83.

SUPPLEMENTARY NOTES

PAGE 28.—The vacuum calorimeter has recently been perfected in various respects in my old Institute by F. Simon and F. Lange, who used it for further very valuable measurements.

F. Simon ("Ann. d. Physik," **68,** p. 241, 1922), by constructing the whole apparatus of metal, secured better dimensions and greater stability; he used a thermo-couple for the measurement of temperature. Of his results I may mention here, as important for our purpose, the experimental confirmation of the Heat Theorem for the formation of cuprous iodide, though the lack of accurate data for the free energy of formation of the compound rendered the confirmation only approximate.

In the same apparatus F. Simon and F. Lange (" Zeitschr. f. Physik," **15,** p. 312, 1923) were able, by a very original artifice, to cool the calorimeter down to 9° abs., and to carry out accurate measurements of specific heat down to this temperature. Some liquid hydrogen was condensed in the calorimeter K (cf. Fig. 4, p. 30) and pumped off through a German-silver capillary, so that it solidified and the calorimeter cooled to the above low temperature; the temperature of the calorimeter thus fell considerably below that of the bath (*ca.* 20° abs.). The relatively intense cooling could be easily obtained in virtue of the excellent thermal insulation of the calorimeter and of its extremely low heat capacity. The evacuation was continued until practically all the hydrogen had evaporated and the residue remaining contributed no appreciable amount to the thermal capacity of the calorimeter. The temperature of the calorimeter could be measured very accurately from the vapour pressure of the hydrogen, more accurately, in fact, than is possible with resistance thermometers or thermo-couples at these temperatures without very special precautions.

The opportunity was taken to measure at the same time the

heat of evaporation as well as other thermal data for hydrogen; this was the first case in which direct experimental evidence was obtained of the maximum of latent heat, which is of such theoretical importance (cf. p. 130).

F. Lange ("Zeitsch. physik. Chem.," **110**, p. 343, 1924), in continuation of the above work, made use of an "adiabatic calorimeter," the temperature of the bath being brought by electric heating to that of the calorimeter in successive steps. It may be useful, for special purposes, to make the calorimeter by this means more completely independent of the temperature of its surroundings. The author was also successful in bringing the lead resistance thermometer into a form in which its resistance was free from the secular changes which hitherto (cf. p. 43) militated against its employment. Among the numerous results of this work may be cited the successful application of the Heat Theorem to the transformation of tin.

Page 72.—The values of the molecular heats of gases have been collected anew by K. Wohl, who has also added measurements of his own.

Chapter IX, page 99.—An interesting case of a condensed system, namely, the affinity of the transformation of amorphous silica (quartz glass) into crystallized quartz, has been worked out by R. Wietzel ("Zeitsch. anorg. Chem.," **116**, p. 71, 1921) in my old laboratory. Since, however, the specific heat of quartz glass falls off so slowly that it was not possible to get anywhere near the region of the T^3-Law, the experimental investigation of this case is not yet complete. This is one of the cases where it is very desirable that accurate measurement of specific heat should be continued down to helium temperatures. This is the more important in that no direct experimental test of the application of the Heat Theorem to amorphous substances has yet been made with satisfactory reliability. The transformation from quartz into cristobalite could, however, be followed with adequate accuracy from the standpoint of the Heat Theorem.

Reference may be made in this connection to the excellent review by Prof. W. Eitel of Königsberg, "Das Nernstsche Wärmetheorem und seine Bedeutung für mineralogisch-geologische Probleme" ("Fortschritte der Mineralogie," Bd. 8, 1923).

Page 142.—The fact that NO really has an extremely high

value of the quotient $\frac{\lambda}{T_0}$ has been recently demonstrated (H. Goldschmidt, "Zeitschr. f. Physik," **20**, p. 159, 1923) in a thorough investigation carried out in my old laboratory. The value found for $\frac{\lambda}{T}$ was 26·1, that of the conventional chemical constant (p. 136) being 4·0. The number in the brackets in the table (p. 143) is reduced from 1·60 to 1·36. The new numbers are a remarkable confirmation of our approximation formula (p. 141).

Page 185.—The experimental check of the value given by the quantum theory for the chemical constant [*] has quite recently been brought into a new stage, which may well be called extremely surprising, by the work of F. Simon ("Zeitschr. physik. Chem.," **110**: Nernst Jubilee volume, p. 572, 1924). The values of the table on page 185 remain essentially unaltered; for H_2 Simon finds, in a special investigation, $C_0 = -1\cdot57 \pm 0\cdot03$, i.e. a complete agreement with theory (cf. "Zeitschr. f. Physik," **15**, p. 307, 1923); this was the result of careful determinations of the vapour pressure and specific heat of solid hydrogen. Fresh measurements for mercury lead to $C_0 = -1\cdot50$, i.e. to a small but apparently distinct deviation from the theoretical value. For argon F. Born ("Ann. der Physik," **69**, p. 473, 1922) obtained the more accurate value $C_0 = -1\cdot61$, very near the theoretical value; this was deduced from careful measurements on well-purified samples, such that the appreciable errors which affect Crommelin's measurements were not operative. Values for the chemical constant which were in each case considerably too high were found by Simon (*loc. cit.*) and by K. Wohl ("Zeitschr. physik. Chem.," **110**, p. 166, 1924) for a number of other substances, all of which, curiously, are fairly high boiling, or at any rate are evolved from the condensate with a very large heat absorption.

In several cases the theoretical value has been obtained, and it is a strange coincidence that these should be the cases in which the experimental material available was many years ago sufficient for the calculation of useful values of the chemical constant.

---

[*] Egerton ("Phil. Mag.," **39**, 1, 1920) gives data for some other elements, with a critique of the then available measurements for mercury.—TR.

This agreement is evidence that Stern's remarkable calculation is at any rate not entirely wide of the mark, but rather that it undoubtedly contains a nucleus of truth.

The variation in specific heat of the substances which function as condensates, in the cases which have been examined at extremely low temperatures, might show some still unknown irregularities, but this possibility is excluded because the deviations would disappear only if the fall in specific heat of the condensate were more rapid; but the measurements have been continued so far that even a sudden diminution to zero below the last measured value would not lead to any essential change in the result.

If, therefore, the deviations are real, the question arises what property of the gas, other than its molecular weight, affects the chemical constant; as Simon argues, there might be forces of cohesion which are still unknown.

If we remember the two very different routes which led to the theoretical calculation of chemical constants (pp. 169 and 205), we must regard it as probable—

(1) That Stern's body should not represent an idealized solid, the decrease in specific heat, for example, taking place according to laws which are not so simple as those supposed.

(2) That the degeneration of gases may show some peculiarities, unknown at present, for various gases.

However this may be, what matters for the purpose considered in this book is not whether we can calculate the chemical constant from molecular theory, but rather whether it exists at all; it is only the latter that our Heat Theorem requires, and of this the investigations of Simon and Wohl, so far from adducing any refutation, have, as the authors themselves remark, given fresh confirmation.

Page 200.—The question of the degeneration of gases cannot yet be regarded as settled; * as shown in the first edition of this book, it is a consequence of the idea that our Heat Theorem is applicable also to gases, and this is definitely fixed if the variation

---

* See the argument of F. A. Lindemann ("Phil. Mag.," **39**, 21, 1920) on the dimensions of the chemical constant.—TR.

of specific heat of a gas at constant volume may be assumed to be known down to the lowest temperatures.

As Dr. Bennewitz informed me, it is convenient to add the further hypothesis that at high temperatures a gas obeys the gas laws quite exactly; only on this assumption is the cyclic process which I have described on page 194 perfectly convincing, for only then can it be supposed that at high temperatures a gas no longer shows the Joule-Thomson effect. This assumption I have taken to be self-evident on page 199, line 24, but it is only very probable, not self-evident. Owing to the smallness of the energy of gases at the absolute zero, the above assumption is hardly capable of experimental proof. In other words, the possibility must be taken into account that the formula at the foot of page 194 does not hold quite exactly for high values of T'.

We assume below that this formula is rigidly accurate; the energy at the absolute zero is then given by a simple application of it as

$$- U_0 = \tfrac{3}{2}RT - E \ (\lim T = \infty),$$

or, as may be readily found,

$$- U_0 = \tfrac{3}{4}Nh\nu.$$

In the text on page 205 the factor $\tfrac{3}{2}$ is given instead of $\tfrac{3}{4}$, which was only due to a mistake, but leads to essentially different formulæ for the degeneration of gases, as was shown by Dr. Bennewitz in a very important study of the degeneration of gases and the energy at absolute zero (Nernst Jubilee volume, " Zeitschr. physik. Chem.," **110**, p. 725, 1924).

Actually equation (**146***b*) now takes the form,

$$- A = \tfrac{3}{2}RT \log_e \left(e^{\tfrac{\beta\nu}{T}} - 1 \right) - \tfrac{3}{4}R\beta\nu.$$

For the pressure we then obtain

$$p = - \left(\tfrac{3}{2}R \cdot \beta \frac{e^{\tfrac{\beta\nu}{T}}}{e^{\tfrac{\beta\nu}{T}} - 1} - \tfrac{3}{4}R\beta \right) \frac{\partial \nu}{\partial V},$$

and, after a few simple rearrangements,

$$p = \tfrac{1}{2}\frac{R\beta\nu}{V} \cdot \frac{e^{\tfrac{\nu\beta}{T}} + 1}{e^{\tfrac{\beta\nu}{T}} - 1}.$$

SUPPLEMENTARY NOTES

As Bennewitz points out (*loc. cit.*, p. 754), the calculations made on page 206 are now substantially changed, in that equation **(150)** assumes the form

$$pV = RT\left\{1 + \tfrac{1}{12}\left(\tfrac{\beta \nu}{T}\right)^2\right\}.$$

The influence of the degeneration of gases is so much smaller that, on the one hand, every point of difference from existing observations completely disappears and, on the other hand, a direct experimental test of the degeneration of gases recedes practically beyond the bounds of possibility. It is also very noteworthy that in equation **(144)** the term after λ_0 entirely disappears. All the above formulæ hold, of course, only on the assumption that the specific heat of gases at low temperatures behaves as is required by the energy formula **(138)**, page 202.

For an attempt at an indirect test of the degeneration of gases, cf. my note, "Ber. d. preuss. Akad.," 1919, p. 118. In the examination of the viscosity of gases at very low temperatures, P. Günther ("Zeitschr. physik. Chem.," **110**, p. 626, 1924) found what was essentially the predicted remarkable behaviour, when he succeeded in measuring the viscosity of hydrogen and helium down to 16° abs.

In one of the numerous thermodynamical papers of Mario Wagner, which, though pertinent in parts to our Heat Theorem, I see no reason to consider here, there occurs the passage ("Zeitschr. physik. Chem.," **96**, p. 289, 1920) :—

"According to the Second Law we must have

$$\int_0^T \frac{1}{T}(C_{v_1} - C_{v_2})dT = R \log_e \frac{v_1}{v_2};$$

hence it follows that, since the right side is positive $(v_1 > v_2)$,

$$C_{v_1} > C_{v_2}.$$

This inequality demands the second fundamental assumption of Nernst."

This conclusion is not admissible, for the above integral may very well still be positive, although the latter inequality be partially untrue, i.e. if in Fig. 20 (p. 193) the two curves cross in places. It is therefore necessary to make the assumption, as I have done, that this does not occur.

Page 231.—As noted in the first edition, a value of 14,570 cals. was published by G. Jones and E. L. Hartmann for the heat of formation of silver iodide, this having been determined electro-chemically (i.e. from the temperature coefficient of the cell $Ag/AgI/I_2$). As this number is fundamental for testing the Heat Theorem in the present case, the measurements in question have been carefully checked and extended by O. Gerth ("Zeitschr. f. Elektrochem.," **27**, p. 287, 1921). The following list contains the values found up to the present by different methods :—

Thermo-chemical.	Electro-chemical.	Calculated from the Heat Theorem.	Observer.
14,990	15,169	15,079	U. Fischer.
15,100	—	15,188	Koref and Braune.
—	14,570	—	Jones and Hartmann.
—	15,150	—	H. S. Taylor.
—	15,158	—	O. Gerth.

The value of Jones and Hartmann is quite discordant; the sources of error present in the work of these authors must have been detected in Gerth's investigation.

Page 245.—The table on page 245 contains the numerical values which were used in the first edition; in the light of more recent measurements the values of the following constants should now be replaced (cf. my " Theor. Chem.," p. 514) by the more accurate results :—

$$h = 6\cdot54 \times 10^{-27}, \quad N = 6\cdot064 \times 10^{-23}, \quad m = 1\cdot662 \times 10^{-24},$$
$$\mu = 9\cdot00 \times 10^{-28}.$$

Hence it follows that

$$k = 1\cdot371 \times 10^{-16}, \quad \beta = 4\cdot77 \times 10^{-11}.$$

For C_0 (p. 186) we thus have, if the pressure is measured in atmos.,

$$C_0 = -1\cdot588.$$

Page 247.—The numbers in this important table have been calculated afresh by W. Zeidler, Grüneisen having previously pointed out to me a few inaccuracies.

Some Remarks on the Use of the Approximation Formulæ.—Equation (**87**), on page 134, gives an approximation formula

SUPPLEMENTARY NOTES 271

which takes note of the influence of specific heats, at any rate in its main features, and hence gives, except in extreme cases, quite a good approximation, as has been many times shown. As concerns gases, our knowledge of specific heat is still, unfortunately, extremely scanty not only at low temperatures, where condensation is troublesome, but also at high temperatures, where dissociation very frequently causes difficulties. An accurate application of the Heat Theorem is often, therefore, impossible, and frequently even equation (87) cannot be used. It is hardly to be expected that this state of affairs will change in the near future.

Since it is only necessary to introduce correction terms to take account of the specific heats, we are, nevertheless, almost always in a position to deduce the approximate position of the equilibrium from a knowledge simply of the thermal changes, even in those cases in which gases participate ; this is done by applying equation (92) of page 137. It is to be noted that, at any rate within a certain interval of temperature, the effect of the specific heat of the gas can be taken into account by the numerical value of C ; this is very simply done by deriving this value from known chemical equilibria.

Reference has already been made on page 137 to the difference between the two approximation formulæ (87) and (92) ; the latter, which is the simpler and thus, naturally, the less accurate, is to be regarded as a generalization and improvement of the rule of le Chatelier and Matignon, in that it is applicable, not only to the equilibrium between a single gas and one or more condensates, but to equilibria of any kind. Moreover, it is considerably more accurate than this rule, because it takes note, at least in some degree, firstly, of the special nature of the gas concerned, by means of the numerical value of C, and, secondly, of the influence of temperature, by means of the term $1.75 \log T$ (cf. the very instructive examples discussed on pp. 150 and 151).

It may perhaps be pointed out that use is frequently not made of the approximation formula (92), even in scientific or technical work, where it might be of value ; the fault may lie in part with an opposition which may not be as clear as it should be as to the limits of serviceability of the formula.

In these circumstances it may be of use to examine more

closely the difference between the two approximation formulæ in the light of a special example, and for this purpose we shall choose the ammonia equilibrium, in which the effect of the specific heats is relatively large. This example is, moreover, particularly suitable to indicate the two rather different points of view which I had to bear in mind when I put forward the approximation formulæ, namely, 1, the checking of my Heat Theorem, and 2, the calculation of chemical equilibria.

The former object was attained by the proposal of conventional chemical constants, so far at least as is possible from the use of approximation formulæ, with which we have frequently to be content in the absence of accurate knowledge of the specific heats. It proved possible to calculate these quantities by different means, independently of chemical equilibria. In numerous cases the result followed, that chemical equilibria with a gaseous phase could be calculated from the heat change of the chemical process, together with quantities which have nothing to do with chemical processes; this was the first fact which settled any doubts in my mind of there being a deep meaning in my Heat Theorem, although approximate calculations only could be employed.

The problem is to be attacked in quite another manner when our concern is simply to calculate chemical equilibria; in this case it will of course be best to derive the chemical constants from actual chemical equilibria. I have naturally worked in both directions in my numerous calculations, of which I have published, of course, only a small fraction. In my publications I have laid less stress on the accurate calculation of equilibria than on the remarkable fact, that quantities like Trouton's coefficient, and in particular certain coefficients in my vapour-pressure formula, bore a close relation to chemical equilibria. Those who are not very practised in thermodynamical calculations will hardly have recognized this distinction, and for this reason a repetition of the calculation of the ammonia equilibrium will be desirable.

As already explained, it is a simple consequence of arithmetic that, strictly speaking, different values of conventional chemical constants must be employed according as formula (87) or (92) is being used; particularly is this difference significant for hydrogen, the conventional chemical constant of which is out of line with

SUPPLEMENTARY NOTES

the others, which all have a value not far removed from 3·1. When I was mainly concerned with formula (**92**), 2·2 was the suitable value for hydrogen; when our knowledge of specific heats had advanced further so that formula (**87**) could be chiefly used, it was an arithmetical necessity that 1·6 should result as the " conventional chemical constant " of hydrogen, in agreement with its vapour-pressure formula.

In order to make the position clear to those who are not very used to such calculations, let us apply formula (**92**), in the light of the figures available in 1906, to those equilibria in which gaseous hydrogen plays a part, viz. to the reactions :—

1. $2H_2 + O_2 = 2H_2O$.
2. $CO + H_2O = H_2 + CO_2$.
3. $H_2 + Cl_2 = 2HCl$.

The reaction $S_2 + 2H_2 = 2H_2S$ must remain unconsidered since here Q' is unknown. Preuner, in his admirable work on this equilibrium (cf. p. 165), was only able to determine the heat change at high temperatures.

I may now show how we have best to operate with formula (**92**), if we wish to determine approximately the chemical equilibrium in the formation of ammonia, without assuming the chemical constant of hydrogen as known; to this end we calculate its value from the above three well-known equilibria. The constants obtained by me in 1907 for the other gases were :—

N_2 2·6, O_2 2·8, CO 3·6, HCl 3·0, Cl_2 3·0, CO_2 3·2, H_2O 3·7, NH_3 3·3.

1. $3 \log_{10} 0 \cdot 000189 - 0 \cdot 30 = - \dfrac{116000}{4 \cdot 57 \times 1480} + 1 \cdot 75 \log_{10} 1480 + I$.

$I = + 0 \cdot 2$; hence $C_0 = 2 \cdot 4$.

2. Here the material was already available in 1907, which led me to the equation proposed on page 143:

$$\log_{10} K' = \dfrac{10200}{4 \cdot 57 T} + 2 \cdot 05;$$

but from this it follows that $2 \cdot 05 = 3 \cdot 6 + 3 \cdot 7 - C_0 - 3 \cdot 2$, so that $C_0 = 2 \cdot 05$.

3. Here we have (p. 143),

$$\log_{10} K' = -\frac{44000}{4\cdot 57 T} - 0\cdot 70$$

so that $-0\cdot 70 = 3\cdot 0 + C_0 - 6\cdot 0$, and $C_0 = 2\cdot 3$.

In my Silliman lecture (1907) I showed, further (p. 115), that the E.M.F. of the oxyhydrogen cell could be calculated very accurately on the assumption that $C_0 = 2\cdot 2$. With $C_0 = 1\cdot 6$ the value comes to $1\cdot 25$ volts (cf. " Theor. Chem.," p. 854), and with $C_0 = 2\cdot 2$ to $1\cdot 236$ volts, while the correct value is $1\cdot 233$.

Further, it may readily be appreciated, in the light of observations mostly made, it is true, after 1907 (cf. List of Literature, Nernst, No. 25), that for the formation of hydriodic acid the result from the use of equation (92) is again $C_0 = 2\cdot 2$.

Let us now, however, calculate C_0 from Haber's figure for the ammonia equilibrium which I considered in 1906 to be much in error :—

$$\log_{10} \frac{0\cdot 325}{1\cdot 2 \times 10^{-4}} = -\frac{12000}{4\cdot 57 \times 1293} + 1\cdot 75 \log_{10} 1293 + \frac{I}{2},$$

whence $I = 0$, and $C_0 = 1\cdot 33$. If, following Haber (cf. " Theor. Chem.," p. 758), we put Q' (per mol NH_3) not 12,000 cals., but only 11,000 cals., we obtain $C_0 = 1\cdot 2$.

We thus find for C_0 in the above three cases, from the material which was at my disposal in 1906, the values $2\cdot 4$, $2\cdot 05$, and $2\cdot 3$, respectively ; for the formation of hydrogen iodide, for which I had reliable material only some while later, C_0 was found to be $2\cdot 2$. For the formation of ammonia, Haber's oldest value gives $1\cdot 33$, if the value for the heat of formation of ammonia which was at that time the best is used, and *ca.* $1\cdot 2$ if the present most probable value is selected.

In any case, there was reason for me in 1906 to doubt the accuracy of Haber's value ; I did not at the time, however, mention any doubt in print, but communicated it to Haber by letter, in order to suggest to him a revision of the tables, which he had drawn up in 1905, for the ammonia equilibrium ; for I was well aware that there is always some degree of uncertainty in the use of an approximation formula. It was not until my Silliman lecture, after I had, jointly with Jellinek, followed the synthesis of ammonia under high pressure by measurement, that

SUPPLEMENTARY NOTES

I pointed out (1907) that the above discrepancy had partly disappeared.

For the purpose of testing my approximation formulæ, and thus indirectly my Heat Theorem, it was of course very important to me to explain this strikingly large difference. My pupil, Dr. Jost, soon furnished determinations of the ammonia equilibrium which, though not final, certainly gave values fairly near the truth ; I myself determined the specific heat of ammonia at high temperatures. I was thus in a position in 1910 (List of Literature, No. 33) to apply equation **(87)** with fair certainty, and to demonstrate that this formula gave an agreement with the position of the ammonia equilibrium which could be called satisfactory.*

We shall now revert once more to approximation formula **(92)** and calculate C_0 from Haber's final figures for the ammonia equilibrium :—

$$T = 1293 \qquad 773 \qquad 473$$
$$C_0 = 1 \cdot 65 \qquad 1 \cdot 8 \qquad 2 \cdot 2 \qquad (Q' = 12{,}000)$$
$$C_0 = 1 \cdot 5 \qquad 1 \cdot 6 \qquad 1 \cdot 9 \qquad (Q' = 11{,}000)$$

* Relying on the fact that I made my calculations of the ammonia equilibrium first (1906) with $C_0 = 2 \cdot 2$, and later (1910) with $C_0 = 1 \cdot 6$, Haber (cf. " Naturwissenschaften," 1922, p. 1046) considers that the deviation of my theoretical calculation from his first detemination should be regarded as " purely fortuitous." I am sorry I cannot agree with this. For if one compares the ammonia equilibrium with other analogous equilibria, as I did in 1906, without making use of any special value of C_0, one must be surprised at Haber's first measurement. Moreover, I made my calculations quite properly with $C_0 = 2 \cdot 2$ when using formula **(92)**, and with $C_0 = 1 \cdot 6$ when using formula **(87)**. I allow that it would have been more correct to draw up two groups of conventional C values, according to the formula in use ; on the other hand, this is a task which is only now of any avail. It is evident, further, that the approximation formula **(92)** was not sufficient for an accurate theoretical calculation. It is impossible, however, to go so far as to deny, as Haber (*loc. cit.*) seeks to do, the historical connection between the investigations on the ammonia equilibrium and my Heat Theorem, even though the marked discrepancy which I noted in 1906 ($C_0 = 1 \cdot 33$ instead of *ca.* $2 \cdot 2$) is to be referred mainly to the effect of the specific heats, and only in a small degree to the inaccuracy of the value for the equilibrium which was used in the calculation.

We see that at high temperatures there are still, even with these numbers, marked deviations of the value of C_0 from those above calculated. We now know that this behaviour may be referred to an unusually large specific-heat effect. It is of importance to have established that at low temperatures the values of C_0 fall in line with those found in the four reactions above discussed; formula (92) works sufficiently well here.

The above calculations and considerations suggest the idea that it will be possible to confer greater accuracy for practical purposes on the very simple approximation formula (92) by tabulating the "chemical constants" of each gas, corresponding with this formula, for different temperatures, and also by taking special note of the effect of the specific heat in some solid condensates; in view of the large amount of observations now available this will usually be possible. Such a procedure might lead to a very convenient and at the same time fairly reliable calculation of chemical equilibria; it is, of course, rendered possible only by the knowledge that the true chemical constants depend, as required by my Heat Theorem, solely on the nature of the gas. If one has in view only the determination of chemical equilibria, one would derive the most suitable values of such chemical constants mainly from chemical equilibria (including sublimation pressures), rather than keep always in mind, as I had to do at first, the simultaneous verification of the Heat Theorem.

All the above considerations should serve as purely practical reflections; the testing of the Heat Theorem on chemical systems with a gaseous phase must, of course, be sought for by means, particularly, of rigid formulæ, and in many cases this has already been done. In order to conclude these considerations with a rigid application, we shall enquire how the position of the ammonia equilibrium may be determined exactly with the aid of the Heat Theorem. The following method is obviously that which is indicated by experiment.

We apply the Heat Theorem to the following equilibrium :—

$$3H_2 + N_2 = 2NH_3.$$
$$\text{solid} \quad \text{solid}$$

As the true chemical constant of hydrogen is accurately known, we require a knowledge only of the specific heats of crystallized nitrogen and ammonia; this is already available in part, even down to very low temperatures. By means of an application of the Second Law simply, we then proceed to higher temperatures, where first nitrogen and then ammonia acquire measureable vapour pressures; the ammonia equilibrium in a homogeneous gaseous phase is then given, and with the aid, again, simply of the Second Law, we can calculate it for temperatures as high as we wish. The specific heats of the three gases are only required above the temperatures at which we introduce them into the calculation. The heat change must of course be known accurately at some one temperature. The only experimental difficulty is to obtain this figure with sufficient accuracy, but this is a difficulty which may well be overcome.

INDEX OF AUTHORS

Adwentowsky, 142.
Avogadro, 64, 66.

Baker, 133.
Behn, 25.
Bennewitz, 268, 269.
Berthelot, 7, 10, 13, 82, 83, 102, 138, 140.
— D., 66, 70, 76, 181, 207, 208.
Bingham, 132.
Bjerrum, 20, 71, 147.
Bodenstein, 140, 151, 165.
Bodländer, 160.
Bohr, 187.
Boltzmann, 57, 171, 240.
Born, 59, 68.
— F., 266.
Boudouard, 157.
Braune, 114, 115, 116, 118, 156, 231, 232.
Brill, 133, 147, 149.
Brönsted, 20, 23, 107.
Brook, 50.
Budde, 147.
Bulle, 42.
Bunsen, 71.
Byk, 235.

Carnot, 3, 83.
Cederberg, 135, 141, 142, 143, 152, 153, 154.
Clapeyron, 83.
Clausius, 3, 83.
Clay, 35, 37, 50.
Clement, 157.
Cohen, 156.
Crommelin, 180, 181, 182, 183.
Curie, 217.
Czapski, 213.
Czukor, 239.

Debye, 14, 37, 51, 52, 59, 61, 63, 64, 65, 69, 82, 83, 95, 96, 108, 187, 219, 234, 245.
Dewar, 25.
Dieselhorst, 46.
Drägert, 84, 109, 111, 235.

Dulong, 54, 57.
Dunant, 165.

Egerton, 266.
Einstein, 56, 58, 63, 65, 68, 94, 161, 163, 237, 239, 240, 241, 245.
Eitel, 265.
Eucken, 25, 29, 37, 43, 44, 48, 51, 52, 53, 61, 68, 71, 74, 82, 180, 181, 182, 183, 193.
Ewald, 26.

Falck, 133.
v. Falkenstein, 19, 22.
Fischer, U., 113, 114, 115, 116, 156, 158, 231, 233.
Forcrand, 8.

Gaede, 29.
Gans, 84, 235.
Gebhardt, 177.
Gerth, 270.
Gibbs, 5.
Goldschmidt, H., 266.
— R., 19.
Grüneisen, 68, 100, 220, 270.
Günther, 48, 53, 65, 115, 269.
Guye, Ph. A., 66, 76.

Haber, 157, 227, 274, 275.
Halla, 165.
Hartmann, 231, 232, 270.
Helmholtz, 2, 3, 8, 10, 18, 211, 212, 231.
Henning, 71.
Hertz, 176, 177.
Heuse, 74.
van't Hoff, 5, 8, 12, 140, 227, 229.
Holborn, 35, 71.
Holland, 165.
Horak, 165.
Horstmann, 7.
Hulett, 233.

Isnardi, 19, 187.

Jahn, 213.
Jewett, 177.

Jones, 231, 232, 270.
Jost, 157, 275.
Jüttner, 236.

KAMERLINGH-ONNES, 35, 36, 37, 50, 51, 52, 61, 68, 135, 234.
v. Kármán, 59, 68.
Katayama, 165, 214.
Keesom, 51, 52, 61, 192, 200, 201, 202, 208, 218, 222, 234.
Keutel, 71.
Kirchhoff, 2, 83.
Knietsch, 15.
Knudsen, 176, 177.
v. Kohner, 165, 183.
Kopp, 57, 62.
Koref, 26, 28, 68, 103, 107, 108, 114, 115, 116, 118, 156, 165, 178, 231, 232.
Kundt, 74.
Kurbatoff, 175.

LANGE, 264, 265.
Langen, 184, 189.
Langmuir, 16, 19, 20, 187.
le Chatelier, 8, 9, 71, 149, 160, 271.
Lemoine, 140.
Levy, 71.
Lewis, 20, 22.
Linde, 33.
Lindemann, F. A., 26, 52, 59, 61, 66, 68, 95, 267.
— Ch. F., 101.
Löwenstein, 16, 20.
Lorentz, H. A., 238, 239.
Lorenz, R., 214.
Lummerzheim, 148.

MADELUNG, 237.
Magnus, 28.
Matignon, 8, 149, 160, 271.
Meyer, V., 16.
Miething, 97, 115, 184, 245.
Miguez, 235.
Morley, 177.
Mündel, 133, 134.

NAUMANN, 133, 158.
Neumann, 57, 62.

VAN OORDT, 157.
Oosterhuis, 217.

PARTINGTON, 71.
Petit, 54, 57.
Pfaundler, 177.
Pier, 17, 18, 20, 71, 147, 188.

Planck, 54, 58, 60, 85, 101, 128, 167, 192, 200, 235, 236, 245.
Pólányi, 92, 193, 237, 239.
Pollitzer, 29, 61, 95, 116, 145, 147, 151, 156, 165, 176, 178.
Preuner, 16, 165.
Pusch, 164.

RAMSAY, 177.
Ratnowsky, 181.
Regnault, 71, 74, 108.
Rhead, 157.
Richards, 227, 228, 229, 230, 231.
Röntgen, 101.
Rubens, 33, 68.
Russel, H., 148.
Russell, A. S., 26.

SACKUR, 169, 174, 175, 192, 205, 208, 209, 236.
Scheel, 74.
Schimank, 43, 68, 131, 180.
Schlesinger, 26.
Schonfliess, 172.
Schottky, 26, 110, 113.
Schwarzschild, 96.
Schwers, 37, 39, 43, 44, 48, 51, 52, 61, 65, 82, 234.
Seibert, 233.
Siegel, 20, 21, 71.
v. Siemens, 44, 133.
Siggel, 110.
Simon, 264, 266, 267.
Sommerfeld, 192, 200, 201.
Stafford, 19.
Stern, 170, 174, 185, 205, 240, 267.
Stock, 35, 44.
Strecker, 71.
Sutherland, 68.

TAMMANN, 24, 103, 104, 105, 107.
Taylor, 232, 233.
Tetrode, 169, 174, 175, 192, 197, 200, 201, 202, 205.
Thibaut, 71.
Thomsen, J., 6, 7, 13, 113, 151, 156, 158, 213.
Thomson, W., 8, 217.
Tilden, 25.
Tower, 18.
Travers, 35, 50, 180.
Trouton, 135, 142, 144.

VOLLER, 71.

VAN DER WAALS, 101, 134, 207, 208.

# INDEX

Wagner, 269.
Warburg, 74, 162, 163, 217.
v. Wartenberg, 16, 19, 20, 22, 23, 145, 147, 158, 165.
Wasjuchnow, 157.
Wheeler, 157.
Wiedemann, 71.
Wietzel, G., 44, 46, 218.
— R., 265.

Wigand, 108, 110.
Winternitz, 150, 183.
Wohl, 71, 265, 266.
Wolff, 165.

YOUNG, 131, 177.

ZEIDLER, 270.

A CATALOGUE OF SELECTED DOVER BOOKS
IN ALL FIELDS OF INTEREST

A CATALOGUE OF SELECTED DOVER BOOKS IN ALL FIELDS OF INTEREST

WHAT IS SCIENCE?, *N. Campbell*
The role of experiment and measurement, the function of mathematics, the nature of scientific laws, the difference between laws and theories, the limitations of science, and many similarly provocative topics are treated clearly and without technicalities by an eminent scientist. "Still an excellent introduction to scientific philosophy," H. Margenau in *Physics Today*. "A first-rate primer ... deserves a wide audience," *Scientific American*. 192pp. 5⅜ x 8.
S43 Paperbound $1.25

THE NATURE OF LIGHT AND COLOUR IN THE OPEN AIR, *M. Minnaert*
Why are shadows sometimes blue, sometimes green, or other colors depending on the light and surroundings? What causes mirages? Why do multiple suns and moons appear in the sky? Professor Minnaert explains these unusual phenomena and hundreds of others in simple, easy-to-understand terms based on optical laws and the properties of light and color. No mathematics is required but artists, scientists, students, and everyone fascinated by these "tricks" of nature will find thousands of useful and amazing pieces of information. Hundreds of observational experiments are suggested which require no special equipment. 200 illustrations; 42 photos. xvi + 362pp. 5⅜ x 8.
T196 Paperbound $2.00

THE STRANGE STORY OF THE QUANTUM, AN ACCOUNT FOR THE GENERAL READER OF THE GROWTH OF IDEAS UNDERLYING OUR PRESENT ATOMIC KNOWLEDGE, *B. Hoffmann*
Presents lucidly and expertly, with barest amount of mathematics, the problems and theories which led to modern quantum physics. Dr. Hoffmann begins with the closing years of the 19th century, when certain trifling discrepancies were noticed, and with illuminating analogies and examples takes you through the brilliant concepts of Planck, Einstein, Pauli, Broglie, Bohr, Schroedinger, Heisenberg, Dirac, Sommerfeld, Feynman, etc. This edition includes a new, long postscript carrying the story through 1958. "Of the books attempting an account of the history and contents of our modern atomic physics which have come to my attention, this is the best," H. Margenau, Yale University, in *American Journal of Physics*. 32 tables and line illustrations. Index. 275pp. 5⅜ x 8.
T518 Paperbound $2.00

GREAT IDEAS OF MODERN MATHEMATICS: THEIR NATURE AND USE, *Jagjit Singh*
Reader with only high school math will understand main mathematical ideas of modern physics, astronomy, genetics, psychology, evolution, etc. better than many who use them as tools, but comprehend little of their basic structure. Author uses his wide knowledge of non-mathematical fields in brilliant exposition of differential equations, matrices, group theory, logic, statistics, problems of mathematical foundations, imaginary numbers, vectors, etc. Original publication. 2 appendixes. 2 indexes. 65 ills. 322pp. 5⅜ x 8.
T587 Paperbound $2.00

CATALOGUE OF DOVER BOOKS

THE RISE OF THE NEW PHYSICS (formerly THE DECLINE OF MECHANISM), *A. d'Abro*
This authoritative and comprehensive 2-volume exposition is unique in scientific publishing. Written for intelligent readers not familiar with higher mathematics, it is the only thorough explanation in non-technical language of modern mathematical-physical theory. Combining both history and exposition, it ranges from classical Newtonian concepts up through the electronic theories of Dirac and Heisenberg, the statistical mechanics of Fermi, and Einstein's relativity theories. "A must for anyone doing serious study in the physical sciences," *J. of Franklin Inst.* 97 illustrations. 991pp. 2 volumes.
T3, T4 Two volume set, paperbound $5.50

THE STRANGE STORY OF THE QUANTUM, AN ACCOUNT FOR THE GENERAL READER OF THE GROWTH OF IDEAS UNDERLYING OUR PRESENT ATOMIC KNOWLEDGE, *B. Hoffmann*
Presents lucidly and expertly, with barest amount of mathematics, the problems and theories which led to modern quantum physics. Dr. Hoffmann begins with the closing years of the 19th century, when certain trifling discrepancies were noticed, and with illuminating analogies and examples takes you through the brilliant concepts of Planck, Einstein, Pauli, de Broglie, Bohr, Schroedinger, Heisenberg, Dirac, Sommerfeld, Feynman, etc. This edition includes a new, long postscript carrying the story through 1958. "Of the books attempting an account of the history and contents of our modern atomic physics which have come to my attention, this is the best," H. Margenau, Yale University, in *American Journal of Physics*. 32 tables and line illustrations. Index. 275pp. $5\frac{3}{8}$ x 8.
T518 Paperbound $2.00

GREAT IDEAS AND THEORIES OF MODERN COSMOLOGY, *Jagjit Singh*
The theories of Jeans, Eddington, Milne, Kant, Bondi, Gold, Newton, Einstein, Gamow, Hoyle, Dirac, Kuiper, Hubble, Weizsäcker and many others on such cosmological questions as the origin of the universe, space and time, planet formation, "continuous creation," the birth, life, and death of the stars, the origin of the galaxies, etc. By the author of the popular *Great Ideas of Modern Mathematics*. A gifted popularizer of science, he makes the most difficult abstractions crystal-clear even to the most non-mathematical reader. Index. xii + 276pp. $5\frac{3}{8}$ x $8\frac{1}{2}$.
T925 Paperbound $2.00

GREAT IDEAS OF MODERN MATHEMATICS: THEIR NATURE AND USE, *Jagjit Singh*
Reader with only high school math will understand main mathematical ideas of modern physics, astronomy, genetics, psychology, evolution, etc., better than many who use them as tools, but comprehend little of their basic structure. Author uses his wide knowledge of non-mathematical fields in brilliant exposition of differential equations, matrices, group theory, logic, statistics, problems of mathematical foundations, imaginary numbers, vectors, etc. Original publications, appendices. indexes. 65 illustr. 322pp. $5\frac{3}{8}$ x 8. T587 Paperbound $2.00

THE MATHEMATICS OF GREAT AMATEURS, *Julian L. Coolidge*
Great discoveries made by poets, theologians, philosophers, artists and other non-mathematicians: Omar Khayyam, Leonardo da Vinci, Albrecht Dürer, John Napier, Pascal, Diderot, Bolzano, etc. Surprising accounts of what can result from a non-professional preoccupation with the oldest of sciences. 56 figures. viii + 211pp. $5\frac{3}{8}$ x $8\frac{1}{2}$.
S1009 Paperbound $2.00

CELESTIAL OBJECTS FOR COMMON TELESCOPES,
Rev. T. W. Webb
Classic handbook for the use and pleasure of the amateur astronomer. Of inestimable aid in locating and identifying thousands of celestial objects. Vol I, The Solar System: discussions of the principle and operation of the telescope, procedures of observations and telescope-photography, spectroscopy, etc., precise location information of sun, moon, planets, meteors. Vol. II, The Stars: alphabetical listing of constellations, information on double stars, clusters, stars with unusual spectra, variables, and nebulae, etc. Nearly 4,000 objects noted. Edited and extensively revised by Margaret W. Mayall, director of the American Assn. of Variable Star Observers. New Index by Mrs. Mayall giving the location of all objects mentioned in the text for Epoch 2000. New Precession Table added. New appendices on the planetary satellites, constellation names and abbreviations, and solar system data. Total of 46 illustrations. Total of xxxix + 606pp. 5⅜ x 8. Vol. 1 Paperbound $2.25, Vol. 2 Paperbound $2.25
The set $4.50

PLANETARY THEORY,
E. W. Brown and C. A. Shook
Provides a clear presentation of basic methods for calculating planetary orbits for today's astronomer. Begins with a careful exposition of specialized mathematical topics essential for handling perturbation theory and then goes on to indicate how most of the previous methods reduce ultimately to two general calculation methods: obtaining expressions either for the coordinates of planetary positions or for the elements which determine the perturbed paths. An example of each is given and worked in detail. Corrected edition. Preface. Appendix. Index. xii + 302pp. 5⅜ x 8½. Paperbound $2.25

STAR NAMES AND THEIR MEANINGS,
Richard Hinckley Allen
An unusual book documenting the various attributions of names to the individual stars over the centuries. Here is a treasure-house of information on a topic not normally delved into even by professional astronomers; provides a fascinating background to the stars in folk-lore, literary references, ancient writings, star catalogs and maps over the centuries. Constellation-by-constellation analysis covers hundreds of stars and other asterisms, including the Pleiades, Hyades, Andromedan Nebula, etc. Introduction. Indices. List of authors and authorities. xx + 563pp. 5⅜ x 8½. Paperbound $2.50

A SHORT HISTORY OF ASTRONOMY, A. Berry
Popular standard work for over 50 years, this thorough and accurate volume covers the science from primitive times to the end of the 19th century. After the Greeks and the Middle Ages, individual chapters analyze Copernicus, Brahe, Galileo, Kepler, and Newton, and the mixed reception of their discoveries. Post-Newtonian achievements are then discussed in unusual detail: Halley, Bradley, Lagrange, Laplace, Herschel, Bessel, etc. 2 Indexes. 104 illustrations, 9 portraits. xxxi + 440pp. 5⅜ x 8. Paperbound $2.75

SOME THEORY OF SAMPLING, W. E. Deming
The purpose of this book is to make sampling techniques understandable to and useable by social scientists, industrial managers, and natural scientists who are finding statistics increasingly part of their work. Over 200 exercises, plus dozens of actual applications. 61 tables. 90 figs. xix + 602pp. 5⅜ x 8½.
Paperbound $3.50

CATALOGUE OF DOVER BOOKS

PRINCIPLES OF STRATIGRAPHY,
A. W. Grabau
Classic of 20th century geology, unmatched in scope and comprehensiveness. Nearly 600 pages cover the structure and origins of every kind of sedimentary, hydrogenic, oceanic, pyroclastic, atmoclastic, hydroclastic, marine hydroclastic, and bioclastic rock; metamorphism; erosion; etc. Includes also the constitution of the atmosphere; morphology of oceans, rivers, glaciers; volcanic activities; faults and earthquakes; and fundamental principles of paleontology (nearly 200 pages). New introduction by Prof. M. Kay, Columbia U. 1277 bibliographical entries. 264 diagrams. Tables, maps, etc. Two volume set. Total of xxxii + 1185pp. 5⅜ x 8. Vol. 1 Paperbound $2.50, Vol. 2 Paperbound $2.50, The set $5.00

SNOW CRYSTALS, *W. A. Bentley and W. J. Humphreys*
Over 200 pages of Bentley's famous microphotographs of snow flakes—the product of painstaking, methodical work at his Jericho, Vermont studio. The pictures, which also include plates of frost, glaze and dew on vegetation, spider webs, windowpanes; sleet; graupel or soft hail, were chosen both for their scientific interest and their aesthetic qualities. The wonder of nature's diversity is exhibited in the intricate, beautiful patterns of the snow flakes. Introductory text by W. J. Humphreys. Selected bibliography. 2,453 illustrations. 224pp. 8 x 10¼. Paperbound $3.25

THE BIRTH AND DEVELOPMENT OF THE GEOLOGICAL SCIENCES,
F. D. Adams
Most thorough history of the earth sciences ever written. Geological thought from earliest times to the end of the 19th century, covering over 300 early thinkers & systems: fossils & their explanation, vulcanists vs. neptunists, figured stones & paleontology, generation of stones, dozens of similar topics. 91 illustrations, including medieval, renaissance woodcuts, etc. Index. 632 footnotes, mostly bibliographical. 511pp. 5⅜ x 8. Paperbound $2.75

ORGANIC CHEMISTRY, *F. C. Whitmore*
The entire subject of organic chemistry for the practicing chemist and the advanced student. Storehouse of facts, theories, processes found elsewhere only in specialized journals. Covers aliphatic compounds (500 pages on the properties and synthetic preparation of hydrocarbons, halides, proteins, ketones, etc.), alicyclic compounds, aromatic compounds, heterocyclic compounds, organophosphorus and organometallic compounds. Methods of synthetic preparation analyzed critically throughout. Includes much of biochemical interest. "The scope of this volume is astonishing," *Industrial and Engineering Chemistry*. 12,000-reference index. 2387-item bibliography. Total of x + 1005pp. 5⅜ x 8. Two volume set, paperbound $4.50

THE PHASE RULE AND ITS APPLICATION,
Alexander Findlay
Covering chemical phenomena of 1, 2, 3, 4, and multiple component systems, this "standard work on the subject" (*Nature*, London), has been completely revised and brought up to date by A. N. Campbell and N. O. Smith. Brand new material has been added on such matters as binary, tertiary liquid equilibria, solid solutions in ternary systems, quinary systems of salts and water. Completely revised to triangular coordinates in ternary systems, clarified graphic representation, solid models, etc. 9th revised edition. Author, subject indexes. 236 figures. 505 footnotes, mostly bibliographic. xii + 494pp. 5⅜ x 8. Paperbound $2.75

A Course in Mathematical Analysis,
Edouard Goursat

Trans. by E. R. Hedrick, O. Dunkel, H. G. Bergmann. Classic study of fundamental material thoroughly treated. Extremely lucid exposition of wide range of subject matter for student with one year of calculus. Vol. 1: Derivatives and differentials, definite integrals, expansions in series, applications to geometry. 52 figures, 556pp. Paperbound $2.50. Vol. 2, Part 1: Functions of a complex variable, conformal representations, doubly periodic functions, natural boundaries, etc. 38 figures, 269pp. Paperbound $1.85. Vol. 2, Part 2: Differential equations, Cauchy-Lipschitz method, nonlinear differential equations, simultaneous equations, etc. 308pp. Paperbound $1.85. Vol. 3, Part 1: Variation of solutions, partial differential equations of the second order. 15 figures, 339pp. Paperbound $3.00. Vol. 3, Part 2: Integral equations, calculus of variations. 13 figures, 389pp. Paperbound $3.00

Planets, Stars and Galaxies,
A. E. Fanning

Descriptive astronomy for beginners: the solar system; neighboring galaxies; seasons; quasars; fly-by results from Mars, Venus, Moon; radio astronomy; etc. all simply explained. Revised up to 1966 by author and Prof. D. H. Menzel, former Director, Harvard College Observatory. 29 photos, 16 figures. 189pp. 5⅜ x 8½. Paperbound $1.50

Great Ideas in Information Theory, Language and Cybernetics,
Jagjit Singh

Winner of Unesco's Kalinga Prize covers language, metalanguages, analog and digital computers, neural systems, work of McCulloch, Pitts, von Neumann, Turing, other important topics. No advanced mathematics needed, yet a full discussion without compromise or distortion. 118 figures. ix + 338pp. 5⅜ x 8½. Paperbound $2.00

Geometric Exercises in Paper Folding,
T. Sundara Row

Regular polygons, circles and other curves can be folded or pricked on paper, then used to demonstrate geometric propositions, work out proofs, set up well-known problems. 89 illustrations, photographs of actually folded sheets. xii + 148pp. 5⅜ x 8½. Paperbound $1.00

Visual Illusions, Their Causes, Characteristics and Applications,
M. Luckiesh

The visual process, the structure of the eye, geometric, perspective illusions, influence of angles, illusions of depth and distance, color illusions, lighting effects, illusions in nature, special uses in painting, decoration, architecture, magic, camouflage. New introduction by W. H. Ittleson covers modern developments in this area. 100 illustrations. xxi + 252pp. 5⅜ x 8. Paperbound $1.50

Atoms and Molecules Simply Explained,
B. C. Saunders and R. E. D. Clark

Introduction to chemical phenomena and their applications: cohesion, particles, crystals, tailoring big molecules, chemist as architect, with applications in radioactivity, color photography, synthetics, biochemistry, polymers, and many other important areas. Non technical. 95 figures. x + 299pp. 5⅜ x 8½. Paperbound $1.50

CATALOGUE OF DOVER BOOKS

THE PRINCIPLES OF ELECTROCHEMISTRY,
D. A. MacInnes
Basic equations for almost every subfield of electrochemistry from first principles, referring at all times to the soundest and most recent theories and results; unusually useful as text or as reference. Covers coulometers and Faraday's Law, electrolytic conductance, the Debye-Hueckel method for the theoretical calculation of activity coefficients, concentration cells, standard electrode potentials, thermodynamic ionization constants, pH, potentiometric titrations, irreversible phenomena. Planck's equation, and much more. 2 indices. Appendix. 585-item bibliography. 137 figures. 94 tables. ii + 478pp. 5⅜ x 8⅜.
Paperbound $2.75

MATHEMATICS OF MODERN ENGINEERING,
E. G. Keller and R. E. Doherty
Written for the Advanced Course in Engineering of the General Electric Corporation, deals with the engineering use of determinants, tensors, the Heaviside operational calculus, dyadics, the calculus of variations, etc. Presents underlying principles fully, but emphasis is on the perennial engineering attack of set-up and solve. Indexes. Over 185 figures and tables. Hundreds of exercises, problems, and worked-out examples. References. Two volume set. Total of xxxiii + 623pp. 5⅜ x 8.   Two volume set, paperbound $3.70

AERODYNAMIC THEORY: A GENERAL REVIEW OF PROGRESS,
William F. Durand, editor-in-chief
A monumental joint effort by the world's leading authorities prepared under a grant of the Guggenheim Fund for the Promotion of Aeronautics. Never equalled for breadth, depth, reliability. Contains discussions of special mathematical topics not usually taught in the engineering or technical courses. Also: an extended two-part treatise on Fluid Mechanics, discussions of aerodynamics of perfect fluids, analyses of experiments with wind tunnels, applied airfoil theory, the nonlifting system of the airplane, the air propeller, hydrodynamics of boats and floats, the aerodynamics of cooling, etc. Contributing experts include Munk, Giacomelli, Prandtl, Toussaint, Von Karman, Klemperer, among others. Unabridged republication. 6 volumes. Total of 1,012 figures, 12 plates, 2,186pp. Bibliographies. Notes. Indices. 5⅜ x 8½.
Six volume set, paperbound $13.50

FUNDAMENTALS OF HYDRO- AND AEROMECHANICS,
L. Prandtl and O. G. Tietjens
The well-known standard work based upon Prandtl's lectures at Goettingen. Wherever possible hydrodynamics theory is referred to practical considerations in hydraulics, with the view of unifying theory and experience. Presentation is extremely clear and though primarily physical, mathematical proofs are rigorous and use vector analysis to a considerable extent. An Engineering Society Monograph, 1934. 186 figures. Index. xvi + 270pp. 5⅜ x 8.
Paperbound $2.00

APPLIED HYDRO- AND AEROMECHANICS,
L. Prandtl and O. G. Tietjens
Presents for the most part methods which will be valuable to engineers. Covers flow in pipes, boundary layers, airfoil theory, entry conditions, turbulent flow in pipes, and the boundary layer, determining drag from measurements of pressure and velocity, etc. Unabridged, unaltered. An Engineering Society Monograph. 1934. Index. 226 figures, 28 photographic plates illustrating flow patterns. xvi + 311pp. 5⅜ x 8.   Paperbound $2.00

CATALOGUE OF DOVER BOOKS

EINSTEIN'S THEORY OF RELATIVITY,
Max Born
Revised edition prepared with the collaboration of Gunther Leibfried and Walter Biem. Steering a middle course between superficial popularizations and complex analyses, a Nobel laureate explains Einstein's theories clearly and with special insight. Easily followed by the layman with a knowledge of high school mathematics, the book has been thoroughly revised and extended to modernize those sections of the well-known original editions which are now out of date. After a comprehensive review of classical physics, Born's discussion of special and general theories of relativity covers such topics as simultaneity, kinematics, Einstein's mechanics and dynamics, relativity of arbitrary motions, the geometry of curved surfaces, the space-time continuum, and many others. Index. Illustrations, vii + 376pp. 5⅜ x 8. Paperbound $2.00

THE PRINCIPLE OF RELATIVITY,
A. Einstein, H. Lorentz, H. Minkowski, H. Weyl
These are the 11 basic papers that founded the general and special theories of relativity, all translated into English. Two papers by Lorentz on the Michelson experiment, electromagnetic phenomena. Minkowski's *Space & Time*, and Weyl's *Gravitation & Electricity*. 7 epoch-making papers by Einstein: *Electromagnetics of Moving Bodies, Influence of Gravitation in Propagation of Light, Cosmological Considerations, General Theory*, and 3 others. 7 diagrams. Special notes by A. Sommerfeld. 224pp. 5⅜ x 8. Paperbound $2.00

ATOMIC SPECTRA AND ATOMIC STRUCTURE,
G. Herzberg
Excellent general survey for chemists, physicists specializing in other fields. Partial contents: simplest line spectra and elements of atomic theory, building-up principle and periodic system of elements, hyperfine structure of spectral lines, some experiments and applications. Bibliography. 80 figures. Index. xii + 257pp. 5⅜ x 8. Paperbound $2.00

PRINCIPLES OF QUANTUM MECHANICS,
W. V. Houston
Enables student with working knowledge of elementary mathematical physics to develop facility in use of quantum mechanics, understand published work in field. Formulates quantum mechanics in terms of Schroedinger's wave mechanics. Studies evidence for quantum theory, for inadequacy of classical mechanics, 2 postulates of quantum mechanics; numerous important, fruitful applications of quantum mechanics in spectroscopy, collision problems, electrons in solids; other topics. "One of the most rewarding features . . . is the interlacing of problems with text," *Amer. J. of Physics*. Corrected edition. 21 illus. Index. 296pp. 5⅜ x 8. Paperbound $2.00

PHYSICAL PRINCIPLES OF THE QUANTUM THEORY,
Werner Heisenberg
A Nobel laureate discusses quantum theory; Heisenberg's own work; Compton, Schroedinger, Wilson, Einstein, many others. Written for physicists, chemists who are not specialists in quantum theory. Only elementary formulae are considered in the text; there is a mathematical appendix for specialists. Profound without sacrifice of clarity. Translated by C. Eckart, F. Hoyt. 18 figures. 192pp. 5⅜ x 8. Paperbound $1.50

CATALOGUE OF DOVER BOOKS

THEORY OF SETS,
E. Kambe
Clearest, amplest introduction in English, well suited for independent study. Subdivision of main theory, such as theory of sets of points, are discussed, but emphasis is on general theory. Partial contents: rudiments of set theory, arbitrary sets and their cardinal numbers, ordered sets and their order types, well-ordered sets and their cardinal numbers. Bibliography. Key to symbols. Index. vii + 144pp. 5⅜ x 8. Paperbound $1.50

ADVANCED CALCULUS,
E. B. Wilson
An unabridged reprinting of the work which continues to be recognized as one of the most comprehensive and useful texts in the field. It contains an immense amount of well-presented, fundamental material, including chapters on vector functions, ordinary differential equations, special functions, calculus of variations, etc., which are excellent introductions to these areas. For students with only one year of calculus, more than 1300 exercises cover both pure math and applications to engineering and physical problems. For engineers, physicists, etc., this work, with its 54 page introductory review, is the ideal reference and refresher. Index. ix + 566pp. 5⅜ x 8. Paperbound $3.00

ELEMENTS OF THE THEORY OF REAL FUNCTIONS,
J. E. Littlewood
Based on lectures given at Trinity College, Cambridge, this book has proved to be extremely successful in introducing graduate students to the modern theory of functions. It offers a full and concise coverage of classes and cardinal numbers, well-ordered series, other types of series, and elements of the theory of sets of points. 3rd revised edition. vii + 71pp. 5⅜ x 8. Paperbound $1.25

INTRODUCTION TO THE THEORY OF FOURIER'S SERIES AND INTEGRALS,
H. S. Carslaw
3rd revised edition. This excellent introduction is an outgrowth of the author's courses at Cambridge. Historical introduction, rational and irrational numbers, infinite sequences and series, functions of a single variable, definite integral, Fourier series, Fourier integrals, and similar topics. Appendices discuss practical harmonic analysis, periodogram analysis. Lebesgue's theory. Indexes. 84 examples, bibliography. xii + 368pp. 5⅜ x 8. Paperbound $2.25

INFINITE SEQUENCES AND SERIES,
Konrad Knopp
First publication in any language! Excellent introduction to 2 topics of modern mathematics, designed to give the student background to penetrate farther by himself. Sequences & sets, real & complex numbers, etc. Functions of a real & complex variable. Sequences & series. Infinite series. Convergent power series. Expansion of elementary functions. Numerical evaluation of series. Bibliography. v + 186pp. 5⅜ x 8. Paperbound $1.85

INTRODUCTION TO THE DIFFERENTIAL EQUATIONS OF PHYSICS,
L. Hopf
Especially valuable to the engineer with no math beyond elementary calculus. Emphasizing intuitive rather than formal aspects of concepts, the author covers an extensive territory. Partial contents: Law of causality, energy theorem, damped oscillations, coupling by friction, cylindrical and spherical coordinates, heat source, etc. Index. 48 figures. 160pp. 5⅜ x 8. Paperbound $1.35

CATALOGUE OF DOVER BOOKS

A SHORT ACCOUNT OF THE HISTORY OF MATHEMATICS,
W. W. Rouse Ball
Last previous edition (1908) hailed by mathematicians and laymen for lucid overview of math as living science, for understandable presentation of individual contributions of great mathematicians. Treats lives, discoveries of every important school and figure from Egypt, Phoenicia to late nineteenth century. Greek schools of Ionia, Cyzicus, Alexandria, Byzantium, Pythagoras; primitive arithmetic; Middle Ages and Renaissance, including European and Asiatic contributions; modern math of Descartes, Pascal, Wallis, Huygens, Newton, Euler, Lambert, Laplace, scores more. More emphasis on historical development, exposition of ideas than other books on subject. Non-technical, readable text can be followed with no more preparation than high-school algebra. Index. 544pp. 5⅜ x 8. Paperbound $2.25

GREAT IDEAS AND THEORIES OF MODERN COSMOLOGY, *Jagjit Singh*
Companion volume to author's popular "Great Ideas of Modern Mathematics" (Dover, $2.00). The best non-technical survey of post-Einstein attempts to answer perhaps unanswerable questions of origin, age of Universe, possibility of life on other worlds, etc. Fundamental theories of cosmology and cosmogony recounted, explained, evaluated in light of most recent data: Einstein's concepts of relativity, space-time; Milne's a priori world-system; astrophysical theories of Jeans, Eddington; Hoyle's "continuous creation;" contributions of dozens more scientists. A faithful, comprehensive critical summary of complex material presented in an extremely well-written text intended for laymen. Original publication. Index. xii + 276pp. 5⅜ x 8½. Paperbound $2.00

THE RESTLESS UNIVERSE, *Max Born*
A remarkably lucid account by a Nobel Laureate of recent theories of wave mechanics, behavior of gases, electrons and ions, waves and particles, electronic structure of the atom, nuclear physics, and similar topics. "Much more thorough and deeper than most attempts... easy and delightful," *Chemical and Engineering News*. Special feature: 7 animated sequences of 60 figures each showing such phenomena as gas molecules in motion, the scattering of alpha particles, etc. 11 full-page plates of photographs. Total of nearly 600 illustrations. 351pp. 6⅛ x 9¼. Paperbound $2.00

PLANETS, STARS AND GALAXIES: DESCRIPTIVE ASTRONOMY FOR BEGINNERS,
A. E. Fanning
What causes the progression of the seasons? Phases of the moon? The Aurora Borealis? How much does the sun weigh? What are the chances of life on our sister planets? Absorbing introduction to astronomy, incorporating the latest discoveries and theories: the solar wind, the surface temperature of Venus, the pock-marked face of Mars, quasars, and much more. Places you on the frontiers of one of the most vital sciences of our time. Revised (1966). Introduction by Donald H. Menzel, Harvard University. References. Index. 45 illustrations. 189pp. 5¼ x 8¼. Paperbound $1.50

GREAT IDEAS IN INFORMATION THEORY, LANGUAGE AND CYBERNETICS,
Jagjit Singh
Non-mathematical, but profound study of information, language, the codes used by men and machines to communicate, the principles of analog and digital computers, work of McCulloch, Pitts, von Neumann, Turing, and Uttley, correspondences between intricate mechanical network of "thinking machines" and more intricate neurophysiological mechanism of human brain. Indexes. 118 figures. 50 tables. ix + 338pp. 5⅜ x 8½. Paperbound $2.00

CATALOGUE OF DOVER BOOKS

THE MUSIC OF THE SPHERES: THE MATERIAL UNIVERSE — FROM ATOM TO QUASAR, SIMPLY EXPLAINED, *Guy Murchie*
Vast compendium of fact, modern concept and theory, observed and calculated data, historical background guides intelligent layman through the material universe. Brilliant exposition of earth's construction, explanations for moon's craters, atmospheric components of Venus and Mars (with data from recent fly-by's), sun spots, sequences of star birth and death, neighboring galaxies, contributions of Galileo, Tycho Brahe, Kepler, etc.; and (Vol. 2) construction of the atom (describing newly discovered sigma and xi subatomic particles), theories of sound, color and light, space and time, including relativity theory, quantum theory, wave theory, probability theory, work of Newton, Maxwell, Faraday, Einstein, de Broglie, etc. "Best presentation yet offered to the intelligent general reader," *Saturday Review*. Revised (1967). Index. 319 illustrations by the author. Total of xx + 644pp. 5⅜ x 8½.
Vol. 1 Paperbound $2.00, Vol. 2 Paperbound $2.00, The set $4.00

FOUR LECTURES ON RELATIVITY AND SPACE, *Charles Proteus Steinmetz*
Lecture series, given by great mathematician and electrical engineer, generally considered one of the best popular-level expositions of special and general relativity theories and related questions. Steinmetz translates complex mathematical reasoning into language accessible to laymen through analogy, example and comparison. Among topics covered are relativity of motion, location, time; of mass; acceleration; 4-dimensional time-space; geometry of the gravitational field; curvature and bending of space; non-Euclidean geometry. Index. 40 illustrations. x + 142pp. 5⅜ x 8½. Paperbound $1.35

HOW TO KNOW THE WILD FLOWERS, *Mrs. William Starr Dana*
Classic nature book that has introduced thousands to wonders of American wild flowers. Color-season principle of organization is easy to use, even by those with no botanical training, and the genial, refreshing discussions of history, folklore, uses of over 1,000 native and escape flowers, foliage plants are informative as well as fun to read. Over 170 full-page plates, collected from several editions, may be colored in to make permanent records of finds. Revised to conform with 1950 edition of Gray's Manual of Botany. xlii + 438pp. 5⅜ x 8½. Paperbound $2.00

MANUAL OF THE TREES OF NORTH AMERICA, *Charles Sprague Sargent*
Still unsurpassed as most comprehensive, reliable study of North American tree characteristics, precise locations and distribution. By dean of American dendrologists. Every tree native to U.S., Canada, Alaska; 185 genera, 717 species, described in detail—leaves, flowers, fruit, winterbuds, bark, wood, growth habits, etc. plus discussion of varieties and local variants, immaturity variations. Over 100 keys, including unusual 11-page analytical key to genera, aid in identification. 783 clear illustrations of flowers, fruit, leaves. An unmatched permanent reference work for all nature lovers. Second enlarged (1926) edition. Synopsis of families. Analytical key to genera. Glossary of technical terms. Index. 783 illustrations, 1 map. Total of 982pp. 5⅜ x 8.
Vol. 1 Paperbound $2.25, Vol. 2 Paperbound $2.25, The set $4.50

IT'S FUN TO MAKE THINGS FROM SCRAP MATERIALS,
Evelyn Glantz Hershoff

What use are empty spools, tin cans, bottle tops? What can be made from rubber bands, clothes pins, paper clips, and buttons? This book provides simply worded instructions and large diagrams showing you how to make cookie cutters, toy trucks, paper turkeys, Halloween masks, telephone sets, aprons, linoleum block- and spatter prints — in all 399 projects! Many are easy enough for young children to figure out for themselves; some challenging enough to entertain adults; all are remarkably ingenious ways to make things from materials that cost pennies or less! Formerly "Scrap Fun for Everyone." Index. 214 illustrations. 373pp. 5⅜ x 8½. Paperbound $1.50

SYMBOLIC LOGIC and THE GAME OF LOGIC, *Lewis Carroll*

"Symbolic Logic" is not concerned with modern symbolic logic, but is instead a collection of over 380 problems posed with charm and imagination, using the syllogism and a fascinating diagrammatic method of drawing conclusions. In "The Game of Logic" Carroll's whimsical imagination devises a logical game played with 2 diagrams and counters (included) to manipulate hundreds of tricky syllogisms. The final section, "Hit or Miss" is a lagniappe of 101 additional puzzles in the delightful Carroll manner. Until this reprint edition, both of these books were rarities costing up to $15 each. Symbolic Logic: Index. xxxi + 199pp. The Game of Logic: 96pp. 2 vols. bound as one. 5⅜ x 8.
Paperbound $2.00

MATHEMATICAL PUZZLES OF SAM LOYD, PART I
selected and edited by M. Gardner

Choice puzzles by the greatest American puzzle creator and innovator. Selected from his famous collection, "Cyclopedia of Puzzles," they retain the unique style and historical flavor of the originals. There are posers based on arithmetic, algebra, probability, game theory, route tracing, topology, counter and sliding block, operations research, geometrical dissection. Includes the famous "14-15" puzzle which was a national craze, and his "Horse of a Different Color" which sold millions of copies. 117 of his most ingenious puzzles in all. 120 line drawings and diagrams. Solutions. Selected references. xx + 167pp. 5⅜ x 8.
Paperbound $1.00

STRING FIGURES AND HOW TO MAKE THEM, *Caroline Furness Jayne*

107 string figures plus variations selected from the best primitive and modern examples developed by Navajo, Apache, pygmies of Africa, Eskimo, in Europe, Australia, China, etc. The most readily understandable, easy-to-follow book in English on perennially popular recreation. Crystal-clear exposition; step-by-step diagrams. Everyone from kindergarten children to adults looking for unusual diversion will be endlessly amused. Index. Bibliography. Introduction by A. C. Haddon. 17 full-page plates, 960 illustrations. xxiii + 401pp. 5⅜ x 8½.
Paperbound $2.00

PAPER FOLDING FOR BEGINNERS, *W. D. Murray and F. J. Rigney*

A delightful introduction to the varied and entertaining Japanese art of origami (paper folding), with a full, crystal-clear text that anticipates every difficulty; over 275 clearly labeled diagrams of all important stages in creation. You get results at each stage, since complex figures are logically developed from simpler ones. 43 different pieces are explained: sailboats, frogs, roosters, etc. 6 photographic plates. 279 diagrams. 95pp. 5⅝ x 8⅜. Paperbound $1.00

CATALOGUE OF DOVER BOOKS

PRINCIPLES OF ART HISTORY,
H. Wölfflin
Analyzing such terms as "baroque," "classic," "neoclassic," "primitive," "picturesque," and 164 different works by artists like Botticelli, van Cleve, Dürer, Hobbema, Holbein, Hals, Rembrandt, Titian, Brueghel, Vermeer, and many others, the author establishes the classifications of art history and style on a firm, concrete basis. This classic of art criticism shows what really occurred between the 14th-century primitives and the sophistication of the 18th century in terms of basic attitudes and philosophies. "A remarkable lesson in the art of seeing," *Sat. Rev. of Literature*. Translated from the 7th German edition. 150 illustrations. 254pp. 6⅛ x 9¼. Paperbound $2.00

PRIMITIVE ART,
Franz Boas
This authoritative and exhaustive work by a great American anthropologist covers the entire gamut of primitive art. Pottery, leatherwork, metal work, stone work, wood, basketry, are treated in detail. Theories of primitive art, historical depth in art history, technical virtuosity, unconscious levels of patterning, symbolism, styles, literature, music, dance, etc. A must book for the interested layman, the anthropologist, artist, handicrafter (hundreds of unusual motifs), and the historian. Over 900 illustrations (50 ceramic vessels, 12 totem poles, etc.). 376pp. 5⅜ x 8. Paperbound $2.25

THE GENTLEMAN AND CABINET MAKER'S DIRECTOR,
Thomas Chippendale
A reprint of the 1762 catalogue of furniture designs that went on to influence generations of English and Colonial and Early Republic American furniture makers. The 200 plates, most of them full-page sized, show Chippendale's designs for French (Louis XV), Gothic, and Chinese-manner chairs, sofas, canopy and dome beds, cornices, chamber organs, cabinets, shaving tables, commodes, picture frames, frets, candle stands, chimney pieces, decorations, etc. The drawings are all elegant and highly detailed; many include construction diagrams and elevations. A supplement of 24 photographs shows surviving pieces of original and Chippendale-style pieces of furniture. Brief biography of Chippendale by N. I. Bienenstock, editor of *Furniture World*. Reproduced from the 1762 edition. 200 plates, plus 19 photographic plates. vi + 249pp. 9⅛ x 12¼. Paperbound $3.50

AMERICAN ANTIQUE FURNITURE: A BOOK FOR AMATEURS,
Edgar G. Miller, Jr.
Standard introduction and practical guide to identification of valuable American antique furniture. 2115 illustrations, mostly photographs taken by the author in 148 private homes, are arranged in chronological order in extensive chapters on chairs, sofas, chests, desks, bedsteads, mirrors, tables, clocks, and other articles. Focus is on furniture accessible to the collector, including simpler pieces and a larger than usual coverage of Empire style. Introductory chapters identify structural elements, characteristics of various styles, how to avoid fakes, etc. "We are frequently asked to name some book on American furniture that will meet the requirements of the novice collector, the beginning dealer, and . . . the general public. . . . We believe Mr. Miller's two volumes more completely satisfy this specification than any other work," *Antiques*. Appendix. Index. Total of vi + 1106pp. 7⅞ x 10¾.
Two volume set, paperbound $7.50

CATALOGUE OF DOVER BOOKS

THE BAD CHILD'S BOOK OF BEASTS, MORE BEASTS FOR WORSE CHILDREN, and A MORAL ALPHABET, *H. Belloc*
Hardly and anthology of humorous verse has appeared in the last 50 years without at least a couple of these famous nonsense verses. But one must see the entire volumes — with all the delightful original illustrations by Sir Basil Blackwood — to appreciate fully Belloc's charming and witty verses that play so subacidly on the platitudes of life and morals that beset his day — and ours. A great humor classic. Three books in one. Total of 157pp. 5⅜ x 8.
Paperbound $1.00

THE DEVIL'S DICTIONARY, *Ambrose Bierce*
Sardonic and irreverent barbs puncturing the pomposities and absurdities of American politics, business, religion, literature, and arts, by the country's greatest satirist in the classic tradition. Epigrammatic as Shaw, piercing as Swift, American as Mark Twain, Will Rogers, and Fred Allen, Bierce will always remain the favorite of a small coterie of enthusiasts, and of writers and speakers whom he supplies with "some of the most gorgeous witticisms of the English language" (H. L. Mencken). Over 1000 entries in alphabetical order. 144pp. 5⅜ x 8.
Paperbound $1.00

THE COMPLETE NONSENSE OF EDWARD LEAR.
This is the only complete edition of this master of gentle madness available at a popular price. *A Book of Nonsense, Nonsense Songs, More Nonsense Songs and Stories* in their entirety with all the old favorites that have delighted children and adults for years. The Dong With A Luminous Nose, The Jumblies, The Owl and the Pussycat, and hundreds of other bits of wonderful nonsense. 214 limericks, 3 sets of Nonsense Botany, 5 Nonsense Alphabets, 546 drawings by Lear himself, and much more. 320pp. 5⅜ x 8.
Paperbound $1.00

THE WIT AND HUMOR OF OSCAR WILDE, *ed. by Alvin Redman*
Wilde at his most brilliant, in 1000 epigrams exposing weaknesses and hypocrisies of "civilized" society. Divided into 49 categories—sin, wealth, women, America, etc.—to aid writers, speakers. Includes excerpts from his trials, books, plays, criticism. Formerly "The Epigrams of Oscar Wilde." Introduction by Vyvyan Holland, Wilde's only living son. Introductory essay by editor. 260pp. 5⅜ x 8.
Paperbound $1.00

A CHILD'S PRIMER OF NATURAL HISTORY, *Oliver Herford*
Scarcely an anthology of whimsy and humor has appeared in the last 50 years without a contribution from Oliver Herford. Yet the works from which these examples are drawn have been almost impossible to obtain! Here at last are Herford's improbable definitions of a menagerie of familiar and weird animals, each verse illustrated by the author's own drawings. 24 drawings in 2 colors; 24 additional drawings. vii + 95pp. 6½ x 6.
Paperbound $1.00

THE BROWNIES: THEIR BOOK, *Palmer Cox*
The book that made the Brownies a household word. Generations of readers have enjoyed the antics, predicaments and adventures of these jovial sprites, who emerge from the forest at night to play or to come to the aid of a deserving human. Delightful illustrations by the author decorate nearly every page. 24 short verse tales with 266 illustrations. 155pp. 6⅝ x 9¼.
Paperbound $1.50

CATALOGUE OF DOVER BOOKS

AN INTRODUCTION TO PATENTS: FOR INVENTORS AND ENGINEERS,
C. D. Tuska
Procedures, precautions, pitfalls and avoiding them; what is patentable; recording; mechanics of interference; ownership and use of patents; infringement. Standard, practical guide, with many concrete situations from actual cases, sample forms, copy of law, etc.; by former Director, Patent Department, R.C.A. 45 illus. 192pp. 5⅜ x 8½.　　　　　　　　　　　　T1169 Paperbound $1.50

LANGUAGE, TRUTH AND LOGIC, *A. J. Ayer*
A clear, careful analysis of the basic ideas of Logical Positivism, building on the work of Schlick, Russell, Carnap and the Viennese School. It covers such topics as the nature of philosophy, science and metaphysics, logic and common sense, and other topics in the philosophy of science. Introduction by Bertrand Russell. 160pp. 5⅜ x 8½.　　　　　　　　　　　　T10 Paperbound $1.35

AN INTELLECTUAL AND CULTURAL HISTORY OF THE WESTERN WORLD,
Harry Elmer Barnes
This exhaustive work by one of America's foremost historians covers every aspect of intellectual life from earliest times to 1965. The arts, sciences, mathematics, philosophy, social institutions, technology, religious institutions, jurisprudence, music, literature, and every other area of cultural history. Scientific portions brought up to date by foremost scholars such as Jagjit Singh, Donald Menzel, etc. Total of 1318pp. 5⅜ x 8½.
　　　　　　　T1275, T1276, T1277　　Three volume set, paperbound $7.50

RECREATIONS IN THE THEORY OF NUMBERS, *Albert H. Beiler*
The number theory behind such diverse topics as perfect numbers, Mersenne numbers, Gauss's congruences, amicable numbers, logarithms, primes, and dozens of other topics, with dozens of problems in the text, and more than 100 teasers in a separate chapter. One of the easiest ways to learn aspects of this important branch of mathematics. 26 figures, 103 tables. vi + 247pp. 5⅜ x 8½.
　　　　　　　　　　　　　　　　　　T1096 Paperbound $2.00

BASIC THEORIES OF PHYSICS, *Peter G. Bergmann*
A thorough coverage of the scientific method and conceptual framework of important topics in classical and modern physics, with concentration on physical ideas. Volume One is concerned with classical mechanics and electrodynamics, including Maxwell's wave equations; volume Two is concerned with heat and quantum theory. Total of xxiii + 580pp. 5⅜ x 8½.
　　　　　　　　　S968, S969　　Two volume set, paperbound $3.85

Prices subject to change without notice.

Available at your book dealer or write for free catalogue to Dept. Adsci, Dover Publications, Inc., 180 Varick St., N.Y., N.Y. 10014. Dover publishes more than 150 books each year on science, elementary and advanced mathematics, biology, music, art, literary history, social sciences and other areas.